100 DAYS

Drive

The Great North American Road Trip

Aaron Lauritsen

100 Days Drive
Copyright © 2016 by Aaron Lauritsen. All rights reserved.

First self-published in 2016.

Photos courtesy of the author.

No part of this book may be used or reproduced in any manner whatsoever without the prior written permission of the publisher, except in the case of brief quotations embodied in reviews.

aarlaur@gmail.com

ISBN 978-1523323777
ISBN 1523323779

Edited by Adria Laycraft

To the influential women who have blessed my life.

Mom for all your sacrifice and hard nights. Shuby for the travel bug and lessons learned. And to my Valley for encouraging me that I should when I didn't think I could.

For all that you do for me my love, and the light you bring, will you make me the luckiest man alive and marry me?

Table of Contents

Leg 1 - The Dream Was Born Of Waking Up Lost   1

Leg 2 - A Drive Through God's Land Alberta   13

Leg 3 - Surf's Up In The Wilds of British Columbia   33

Leg 4 - The Oregon Coast And Border Delays   53

Leg 5 - Dogtown And California Bound   71

Leg 6 - The Sandy Coulee Empire Of The Southwest   89

Leg 7 - Outlaws Of New Mexico   107

Leg 8 - Texas The Lone Star State   119

Leg 9 - New Years In The Big Easy   133

Leg 10 - Florida Sky's Away!   145

Leg 11 – Appalachia   163

Leg 12 - "I Have a Dream" In Washington D.C.   179

Leg 13 - "Pickett's Charge" In Pennsylvania   191

Leg 14 - Live From New York, It's Saturday Night!   211

Leg 15 - New England Charm   231

Leg 16 - Coming Of Age In New Brunswick	243
Leg 17 - Nova Scotia's Ocean Playground	263
Leg 18 - La Belle Province, Quebec	277
Leg 19 - In The Beginning There Was Ontario	291
Leg 20 - Midwest, It's Time To Man Up	303
Leg 21 - The Great Plains Are A Sight To See	319
Leg 22 - Full Circle In The Rangeland	333

Leg 1
The Dream Was Born Of Waking Up Lost

"If you don't know where you are going, any road will get you there." - Lewis Carroll

For millions of people around the world the opportunity to explore a continent by road is a dream that, because of geopolitical forces or instability, is often just out of reach. For those of us who live in North America, where roads transcend regions, cities and borders, the experience is a rite of passage, an institution and a fabric of our very existence.

This is a story with that premise, and as such is perhaps the most magnificent expedition ever chronicled. A journey so life altering, and so epic, that even trailblazers like Lewis & Clark would be envious of the achievements that followed. Okay, so a bit dramatic on my part, but it is the true story of one man's adventure of a lifetime after he puts everything on hold to chase the unmistakable freedom and randomness promised by each bend of the open road. This is a tale of "The Great North American Road Trip!"

Oh, and that man I speak so fondly of? He is none other than the intrepid explorer Aaron Lauritsen. Well, aspiring explorer that is, and I wouldn't be so offended if you've never heard of him, or don't recognize his foot print in time, because he is actually just plain old me, the architect of the Trip and writer of this book.

In fact, I am no adventurer at all; I'm more of a poser who loves the past and tends to live vicariously through some of history's most influential figures, gravitating more towards those who were on the "risk taker" side versus the academic. To understand how this jaunt came about, let me allow you to peer into my life and scrutinize my past as I attempt to explain who I am today, where I was, and where I wanna be.

I was born in Calgary, Alberta, Canada on September 21st 1978, the first day of fall, and for years after my mother Jude would jokingly remind me that, "It was the day the leaves fell." In my defence it happened after the season's first frost, but in all likeliness she was hinting at dubious times ahead. As a kid I was an average student, initially anyways, until junior and senior high when I discovered girls exist. My studies kinda went south from there and moving forward it would be years before I could appreciate a good study sesh or see the value in a book filled rainy night.

Goal driven in different ways, I had a nomadic spirit and was just fortunate enough to scuttle along my path nonchalantly like a gypsy. Planning didn't consume a single thought back then, and for years in fact I preached the creed that, "If you don't know where you're going in life, any road will get you there." I was restless, rebellious and a bit mischievous in my pursuits, and so grew to exhaust my poor mother with an adolescence that saw me constantly sneaking out or being picked up by the police for petty crimes like trespassing to swim in someone's backyard pool or pond.

She was a single parent of three, full time nurse and part time everything in between. As a single mom she did whatever she had to do to pay her dues, and I can only imagine the burden she inherited by marrying the wrong man who was also an abusive deadbeat father to her kids. Yet despite those cards dealt, somehow she found the gumption to carry on. Her story of tribulation is not unique, but it has always been an inspiration for me. And even now, after watching her struggle all those years, I still feel like I've never had a hard day and wonder if I ever will.

Like me, distant realms captivated her and although she would never have the chance to travel, she did from time to time confide that maybe one day she would like to. Until then, her focus was steadfast and her resolve would be set in stone with the priority being to see her children graduate High School. A modest feat for most families, but a monumental task for a single mother of three earning wages below the poverty line.

In life however there are ironies, and so in a twist of fate not long after seeing her children succeed at earning those diplomas, she was diagnosed with breast cancer and died soon thereafter. It wasn't until then that I realized how much she meant to me, I love her very much still, and her guidance has been sorely missed.

As for myself, I was determined not to fall into the same trap of merely surviving day to day. I remember that around age seventeen I picked up a peculiar habit to ensure I was on track to become the worldly man I desired to be, and it's one that has stayed with me: I became a person of lists. I would write lists, rewrite lists, and write lists again to almanac everything from daily errands to reminders of quotes, values, lessons, expectations, challenges and even interpretations. A strange practice for a kid who could barely concentrate.

Especially concerning then was the length of my growing bucket list, already long by most teenage measures. I feared that as the list grew, there wouldn't be enough time to tackle everything, because although some goals were small, others were monumental. I recall my initial bucket list had dozens of bullets; goals of sailing the Seven Sea's, becoming a mountaineer, running marathons, cycling across Canada, being a farmer, tempting fate with skydives, owning a brewery and being the coolest dad ever to name a few. Constantly I pondered with a sense of urgency, "If I don't start now....then when will I be able too?"

As senior year came to a close, I was able to look back and reflect in an attempt to salvage something positive from the school experience, but successes were hard to find. I had terrible marks from skipping more classes then I attended and by year three I had been kicked out several times. Not the kind of academic prowess Mom was hoping for of course, but I did by some small miracle manage to gather enough credits to graduate with my peers. So Jude was over the moon, and in turn I was just happy not to disappoint her.

My education behind me now, and not wanting to attend university, it was finally time to see the world. Shortly, I was sure, I would be in exotic lands, meeting new and exciting characters and contributing in some small way to the fabric of discovery. That all sounds really easy I know, but to do so I needed some serious dinero and a bit of gusto, both of which were commodities lacking in this "Walter Mitty."

Not deterred, and still wanting to leave home, it would come as no surprise to many that at age nineteen I joined the Army. It wasn't a last resort by any means, and though it seemed like an act of desperation, service was actually the realization of another bullet on my bucket list. I had considered the Military option heavily throughout the high school years, but had always shied away from the certain responsibilities involved. I wanted the title of being a professional soldier, but secretly had trepidation about the early morning runs, structured environment and serious people. In short, it all seemed like a lot of hard work, and initially the mere thought was daunting enough.

Getting into the Canadian Army in the late 90's was a competitive and time intensive process. The organization was known to be very professional and well-paying among Armed Forces so it attracted applicants from across the country and around the Commonwealth. Add to that Government cuts, as well as a reduction in the Force after the Cold War, and it took over a year of testing and interviews before I passed recruit selection and was offered a position.

By the time I arrived in Montreal for boot camp I was a very different person then I was a year earlier when I had reluctantly applied. The lengthy selection had given me a chance to get excited about this next chapter and somehow helped me to repress the anxieties I had. I was a bit nervous of course, but as with so many other challenges in life it was the fear of failure that was on my side. And actually once the routine of training began I enjoyed the environment and had no problems with Basic Training, finishing second in my class at the Infantry Battle School. I also took with me an award for marksmanship.

I reflect back on those early years in the Army and laugh to myself now about concerns I had that the world would become a better place and that I would miss out on the action I was seeking. Fortunately or unfortunately perhaps, it never did and I would spend nearly the next decade globetrotting with one course, exercise, tasking or deployment after the other.

It wasn't carefree travel like the kind I had dreamt about, or that my mates who backpacked around Australia bragged about, but it was rewarding nonetheless and in a most intimate way. The work took me to far-off lands, put thousands of miles of soil beneath the soles of my feet and introduced me to a side of humanity most never see. In hindsight I was blessed by it.

The military has a way of building character and instilling confidence in its members. Strong personal qualities that are tested and developed through years of adversity become commonplace. But working overseas made me feel weary and pessimistic. I was beginning to wane. The youthful enthusiasm I had in years past was now just the distant memory of a naive kid. For the first time in a long time, I wanted to go home to seek a life without noise.

What I call 'noise' is the stress, anxiety and pressure of our surroundings. It is simply distraction. For most, noise is the sound of the phone ringing off the hook with bill collectors or telemarketers. It's the constant list of things to do in the back of our minds that never seem subdued, and the feeling of dread we get from a job we don't like to do. For me noise was not having control of my life or of my destiny while I was serving in the Army.

My mind made up, I took my experiences, signed a release, packed my then wife Tammy, our two dogs and my 87' Toyota pickup, called Ruby-Red, then headed West on a quest for a new beginning in the Alberta I loved. We arrived in Calgary to find it wasn't quite as I remembered it to be. It was booming now and bustled feverishly with people from all walks. They flocked to the community for the same reasons I returned, to seek stable work and enjoy a good quality of life in the shadow of the Rocky Mountains.

In less than a generation "Cowtown" had grown substantially. The core that had been a cluster of concrete buildings designed and built in the 1970's was now transformed into towering glass skyscrapers. Signs of that metropolis were everywhere in the newer and faster roads, shopping districts, restaurants and residential developments bursting at the seams of this once modest city. It wasn't long before I was swept up in the euphoria of it all, and intent on seizing my own piece of the pie.

The chronic exhaustion I had felt after my service faded, and once again I felt bullish about the future. Within a year I had bought a new truck to replace my beloved Ruby (who got smashed in a wreck,) purchased a new home in the suburbs, and started what would grow into a successful landscape construction outfit. I found many of my army skills were transferable, and with elbow grease to soften the curve I began to aggressively outcompete the competition and build my business.

Our projects grew in scale as fast as our dump trucks could keep up and recognition in national magazine spreads to showcase award winning work came easier with each season. Professionally things were on track, and initially at least my marriage was benefiting from that. Tammy and I were enjoying more excesses than we were accustomed to and had higher and higher incomes coming in to keep up with the Jones's too.

On the outside looking in, we probably appeared to have a perfect life and the perfect marriage. After all we had everything we had always wanted; security, cash, toys, a house and a great little business. In reality though the picture was bleak and very different than the one we painted for family and friends.

The stress of running a business together had taken a serious toll on our relationship. In the past it was the struggle of getting by that bound us, but without the same woes we began to drift. She and I had spent so much time together over those last years that we no longer appreciated one another. Like many other couples in our circle we didn't even realize our marriage was falling apart until it was long over. The unbreakable bond we had in our 20's, which was constantly tested and rekindled by experiences like me stepping off the plane into her waiting arms after a long deployment, was all but dead and gone.

With our best years of union now behind us and divorce imminent, we set out to openly divvy assets in an attempt to save the friendship. I was determined to buy out her equity in the home and business, so began working long hours installing bricks, troweling out concrete, building irrigation systems and laying miles of sod to clear my debt so we could both put the issue behind us.

I knew the divorce would be taxing, but I hadn't anticipated the toll it would take on me emotionally. The long days and sleepless nights were starting to wear and once again I felt the noise of exhaustion creeping in. I realized I wasn't happy and reckoned that had been the case for a long time. Somewhere along the way, and aided by the indulgences of every day, I had lost my way and found instead the dull existence of an ordinary guy. Don't get me wrong, the comfort of being 'Joe Shmoe' is fine and dandy if that's your goal, but for me it was a scary proposition given that for years I had prescribed to a boundless existence like Captain Cook when he mapped the Pacific.

I remembered how driven I once was to climb every mountain and row every sea. And although I didn't have two cents to rub together during that youthful time, I had felt much more content.

Now creeping up on the wrong side of 40, and a little wiser, I could sense a newfound zest. I wanted my mojo back.

Dreaming non-stop now, I planned a year of travel and discovery to liberate the noise inside of me. I began by selling my business and all my possessions, then found a tenant to live in my house while I was gone. Soon there was nothing holding me back, except a bit of age induced apprehension to warn that grown-ups don't do such ridiculous things as selling their assets to travel the globe.

That was easily bypassed though by starting the year off right on a tropical foot. I did a couple of all-inclusive resorts in Hawaii, visiting Kona, Hilo, Honolulu and Maui. I enjoyed the extreme side of hiking lava fields, snorkeling reefs, diving from cliffy waterfalls and watching Humpback Whales breach. But the ordinary wasn't quite what I needed.

Then I was off to California, Mexico and British Columbia for wine tours before heading north to a remote fishing cabin. Up there in the Lake Country I captured stillness so quiet it hurt, but I didn't feel the contentedness yet. It was savored travel for sure, but predictable and not the authentic vagabond experience I longed for.

Then it dawned on me: when I was twelve, my good friends Shuby and Murray, who are like parents to me, took me on an eye opening three week road trip from Calgary to Toronto. It was a magnificent excursion. We camped out along the way, visited idyllic locations like Devil's Tower and Mount Rushmore and hit Detroit, Chicago, Milwaukee and other carnal cities that, until then, I had only ever heard of in movies. The drive was the first time I left the Prairies, so it had a lasting impact and to this day remains one of my fondest memories. Now an adult and much more appreciative of the small things in life, the idea of a road trip on a much larger scale once again called to me. As I pondered it for a few months, I became convinced that this kind of raw travel, which would be off the beaten track while living out of my truck was just the sort of escape I had been searching for.

The goal now was to conclude the year with a voyage to end all voyages, and the plan was simple; leave everything behind in search of North America's iconic places, spaces and people. I wouldn't plan a single thing, but instead just follow any road that pulled me in.

So with my little black dog Athena-Bear, I threw my backpack and camping gear into the bed of my Toyota truck and headed out the door with the mantra to "explore, experience, then push beyond" in search of an adventure of a lifetime. As I drove away I could see the sun setting west of the Calgary skyline and felt as it did my problems disappear in the rear view mirror with it. It was November 12th of 2013 and I was on my way.

Leg 2

A Drive Through God's Land Alberta

"Home is where one starts from." - T.S. Eliot

My first leg of the trip was a short three hour trek to the provincial capital of Alberta, Edmonton. The intent was to ground myself with familiarity before I searched for roads unknown by doing something I hadn't done in years, reconnect with family and friends via a drive through the Alberta of my youth. In the past the stretch had seemed long and arduous like a chore, but this day it felt sweet and it wasn't long before I found the stride of introspect that would carry me forward the entire journey. Once in that state I began to find humour in the little things that only a week prior would have pissed me off.

Case in point, the day before I left those good friends, Shuby and Murray, treated me to one last Chinese dinner to discuss the trip. The outing had become the norm of late and it was a good excuse to gorge on greasy food while also taking in some good counsel from two seasoned travelers.

Shub and my mother were best friends going back decades. They shared considerable experiences together, ones that powered their relationship and ensured our bond was strong. Years after my Mom's death Shuby confided that the two were soul sisters in life and their paths were destined to cross. For as long as I can remember she has carried the torch of guardian for me, so when she talks I tend to listen.

Over the meal we came up with some interesting scenarios for how the voyage might unfold and discussed potential icons worth a visit. Although the conversation was upbeat and positive, I could sense a bit of angst among us. Shub has spent most of her life in security and over those years had almost certainly developed a sense of caution for the world beyond our borders. Violence on the evening news is nothing new, but gun violence, gangs and the frequent robberies in the States was on a totally different level, and she reminded me to be careful. "Yes dear," I answered back to appease her.

After dinner we all grabbed a fortune cookie. I hastily unwrapped mine, cracked the shell open and began to eat the cookie without regard for etiquette or the message inside. Somewhat annoyed by this she scolded sharply, "Don't you know you're supposed to read the fortune first, you dimwit?"

How cordial, I thought, ashamed of the spotlight she cast on my indiscretion. Looking at her I shrugged as if to half-ass apologize, then reached down for the fortune, picked it up and read it quietly. Smiling ear to ear at what it said, I passed it across the table to let her in on the secret and soon she was smiling too. The fortune, perhaps suggestive of a higher power working to put our minds at ease, read only, "Now is a good time for you to explore."

The timing couldn't have been better, and with a burst of laughter from the table and a strong embrace after, we concluded dinner and went our separate ways. It seemed a good omen to start the journey, and save for one small embarrassing thing not long after it might have been the perfect getaway.

My truck was packed and ready to go for the next day but I wanted some road snacks and a case of beer to put in the cooler, so I headed to the grocery store. At the entrance to the parking lot there was an overhead barrier set up to stop big trucks. It read, "MAX 7 FEET OF CLEARANCE." And just so no one was ever confused, management had affixed another bright yellow fluorescent sign to warn it was there. I'm not sure how anyone could have missed them both....

I drive a little Toyota Tacoma I call "Taco" (yes, very original I know), and on any other given day seven feet would be plenty, even with the lift kit and bed topper I had installed. But on this fateful eve it made perfect sense to stand my four and a half foot mountain bike up on the roof rack...then forget it was there.

Long story short, I punched the gas to get out of the intersection and my bike collided with the clearly marked barrier and created a clamour loud enough to alert every bystander. Somehow I raised the ire of a little old blue hair. The elderly lady had witnessed the entire melee from the other side of the street and was now kind enough to stop traffic with her walker to cross over, then cleverly point out I had hit the sign…..so nice of her.

Frustrated with myself already, I wondered if she was trying to be a Good Samaritan, or if perhaps it was all meant as a slight jab at my misfortune. I'll never know, but being a person who treasures awkwardness, I smiled to thank her while in my head I was thinking, *"Okay lady, I got this! There's nothing to see here. Get off the road already!"*

Fortunately the barrier wasn't damaged at all and the incident only cost me a swath of dignity, but it also begged the question, "why did I bring this bike again?" Years earlier I had spent a month or so touring around Nova Scotia on two wheels and had hoped to duplicate that great junket with another that was similar on this trip through America's Southwest Desert, but with a bent rim now, and not willing to make the effort to get it fixed, I abandoned the ambition as though it were a sign from above (literally), thanked Mother Teresa, then dropped the heap off before I left.

The scenery en route to Edmonton was magnificent and breathtaking. Well defined layers of near, middle and far ground were divvied among endless fields of dormant gold spread across rolling hills that looked like pillows. In the background was the imposing silhouette of the Rockies and I beamed at how much this land had always meant to me.

When I was young, abroad and missing home, I used to close my eyes to blink but would instead be greeted by flashes of this place and what I could only describe to friends I worked with as "God's Land," an awe-inspiring kingdom of mountains, badlands, forests and plains that if you saw them with your own eyes you might agree, I wasn't exaggerating. It was a love affair that had escaped me in recent years, but lost in it as I drove, I felt compelled to raise it back onto the podium it deserved.

I remembered how I used to cut through farmer's fields on the edge of town. Shirtless, with the hot summer sun on my back and my hands palm down, I would caress the tease of swaying crops as warm winds sent tidal waves of energy barreling across to create a rhythm so mesmerizing, the ocean would be envious of its effect. The drive felt like a smug way to say, "Don't ever take this for granted again, Aaron..."

When I arrived in the "City of Champions" I texted my old army buddy, Jason Stoneham or "Stoney" to his friends, to see if he was going to be around for a couple of beers later on. He answered back quickly with "yup" and I sauntered on over to the base where he still lives. Once there I was stricken by the familiarity of the camp and was flooded with memories of a life once lived. It had been a spell since we last met, but I knew that once we were together again we would pick up where we left off.

Stoney welcomed me with a beer in hand and a big grin. He looked and acted like the same person, yet he was different in many ways. His ability to tell a good story still dominated a conversation and his chuckle still filled the room, but he wasn't the tall gangly lad I had known in youth. His body had filled out, and he sported a moustache on his face, the kind we used to laugh at in defiance of the older guys.

To celebrate the reunion we made plans to go to "Dango's" later, a Western bar on the North Side where he moonlights as a bouncer, and where some of his work buddies hang out. While there I met them, all great fella's, and the same approachable blue collared types I remembered army chums to be. As the tab got longer we showed off our best cowboy and attempted square dancing (unsuccessfully), then let loose with stories of our past exploits together.

Military people tend to play tricks on one another to kill boredom when there are lulls in activity. It's a trait common to all service people and it's one you learn to accept, even if you're the target of those pranks.

I began, at Stoney's expense, to recant long forgotten memories stirred by his laughter. I shared with the group a time I emptied all of the air from his tires one Friday afternoon so I could enjoy, from afar, the toil of him filling them up again with a painfully slow pump powered by his cigarette lighter. I enlightened on how on tour he got the nickname "Blue Angel" for his uncanny ability to lie on his back, point his heels to the sky, grab a lighter and with skill ignite his own farts into an impressive blue ball. It was a good catch up and the stories continued unabated throughout the night, Stoney dishing just as much as he took, and both of us telling grander fables with each we round we drank.

A bit hung-over the next morning, I was eager to keep moving but figured it'll have to wait until I at least deal payback first. The night prior I made the mistake of falling asleep on Stoney's couch and awoke with what appeared to be used condoms across my chest. To this day I'm confident they only appeared used, but I'll never know because he doesn't reveal secrets.

In any case it was an eye for an eye now, and so I waited patiently for him to leave the house by pretending I was still sleeping. As soon as I heard the door slam I put my plan in motion. He walks to work each morning, which favoured me because it meant his new F-150 was parked outside in the driveway. Vulnerable, it was the perfect target to be rubberized and needless to say when I was done, folks passing by weren't applauding because they thought he was getting married that day. I even secured a couple of condoms to his exhaust pipe that exploded when he turned the ignition to on, then dangled for some time until an innocent bystander approached to point out how offensive they were.

It only took a few hours for him to realize I got him back and when he did he sent a text that read, "Touché brother." I laughed and fired one back, "Thanks for letting me crash," then made my way west along the Yellowhead Highway towards family, and deeper down memory lane.

Everyone has a place in the back of their minds where as a child they sought safety and refuge with comfortable surroundings. For me that place was my grandfather's farm up in the North Country. Built with his own two hands on a little hill at the end of a gravel road that overlooked his and Grandma's homestead, it was little more than a shack with a battered roof. It had no siding, squeaky floors, curtains for closets and a wood stove. Yet despite its lack of amenities that home had more charm than Buckingham Palace.

The wood veneered walls inside were adorned with artifacts of significance to him, scattered monuments of a full life lived. Deer racks, rifles, buckles and pictures of his two boys riding broncos hung there. It was a place of gatherings, of good times and of bad, and if the ceilings could talk they no doubt would tell endless tales of a home filled with laughter and unconditional love.

As a boy that farm had mystery and excitement that was alien to us kids from the big city. It was an immortal place where children still sat perched in the bed of pickup trucks, and where dogs trailed them as uncles raced down dusty roads in search of livestock. There me, my brother, and the cousins, straddled the edge of life and death, or so we thought, and would issue countless "triple dawg dares" to that effect. So sacred was the dare in fact that with each one tackled, another was sure to follow in a constant game of brinkmanship.

I can remember at about age ten, all of us kids risked life and limb to test our mettle by jumping from the barn roof onto the back of an unsuspecting bull. The challenge of course was to hang on for eight seconds as we had watched the uncles do a thousand times. But in our case that bull was Murray, an undersized steer with a friendly demeanor, the speed of a tortoise and the patience of a hibernating bear. Hardly comparable to the foot stomping beasts of the Calgary Stampede, and even though he never flinched when we landed on him, he looked menacing.

Just as the North had appealed to me for all those years, likely it was the same sense of adventure that drew my grandfather in decades earlier, though I'm sure for very different reasons. In the 1970's he gambled all he knew to move his family there from southern Alberta in search of untamed land and a place to call his own. In every sense he was a true Alberta cowboy, and one of the last of his kind. He was a big man, with big hands, and an imposing physical presence softened by an inviting smile, a straight beaked cap that sat high on his head, and a hand rolled cigarette always smouldering away.

He not only looked the part of the paragon "Marlboro Man" from ads of days long gone, he was that man. His character was larger than life and he was a constant reminder that at one time a man with strong work ethic, masculinity and true grit appealed more to people than Hollywood actors who get paid millions to impersonate them. To neighbors he was known as a horse whisperer for his ability to heal and tame wild animals. To family and friends he was known as Dad or as Moose. But to me he was just plain old Gramps, and having no other male figures in my life, he was my shining light. In Fact until I was about sixteen, if you would have asked, "Who hung the Moon, Aaron?" I would have almost certainly answered back, "Grandpa did."

He died many years ago when I was still in the Army and posted out East. Yet despite the distance created by his passing I still think of him daily, and often wonder what it would have been like to meet him in his prime or to be a fly on the wall to hear his laughter one last time. Towards the end of his life and far removed from the strong trail days of his youth, the old cowboy adored by all just seemed to fade away. Lost in contemplation, he would sit for hours on end staring through the front window of his home to watch the apples in his yard fall. No doubt reflecting on his own journey and perhaps even accepting his time was coming to an end.

    His loss had a profound effect on me and it's safe to say that I still praise the influences he gave. Even as he neared that end I never accepted he would die. Naive, I figured he would be okay, pull through, and live forever just by willing the illness away. That was the hope anyways, and as I've learned in life since, even hope with its powerful ability to lift spirits, raise optimism, and curve the tides of despair, cannot turn back the hands of time.

    Mom loved the open road too. As I continued towards Whitecourt often I caught glimpses of myself smiling in the rear view mirror as I passed haunts where she would stop en route for a pack of smokes to combat the stress of bickering tots who posed the same question over and over, "Are we there yet?"

    She used to pile us last minute into the back of an old beat-up Buick that she bought used for the bargain bottom price of three hundred dollars. Painted mat brown to conceal the rust, and with a torn roof panel to pretty it up, the clunker made for an interesting outline of a once proud behemoth sailing down the Queen Elizabeth 2 Highway. Mom being a strong, independent type though, she would force us to swallow our pride by reminding us, "Kids...it's not much, but it's all we got." Idle words though for a lot who wished its demise, especially after each push start of the engine. At each service station we were required to push until the motor roared to life, then rush to get back in.

I think back and recognize with gratitude that her ability to constantly adapt, innovate and multi-task had a lasting and positive impact. The careful consideration she had to put into the simplest of daily tasks was complex to say the least but understandably common to many single mothers who are forced to make due. For years she drove that rickety car, and every time she shut it off she had to ensure it was parked on a hill or in an area where she could get a good running start. Despite daily challenges like these she never complained but instead carried herself with quiet dignity. The message was clear and resonates to this day with another mother I hold dear who says, "Make hay when the sun shines."

It was late afternoon when I rolled into Whitecourt. I knew everybody was still working, at least for a couple of hours yet, so I decided to go for a run before dinner to unwind after the drive. I've been running forever, but I understand not all value the pastime. For those of us who do however the activity is a therapeutic pursuit that offers glimpses of euphoria attainable only when we accomplish a set goal. The sport acts as an escape. One that allows us to root ourselves and rid our bodies, minds and souls of everyday noise by focusing energy instead on the task at hand and the larger picture of health, happiness and the kind of contentedness that comes with chasing the fountain of youth.

Runners tend to have a distinct set of values and a unique culture that appeals to me and separates them from other groups I see. Typically they are goal orientated, positive individuals who are humble in nature and always willing to tackle a new challenge or give a lift to those in need. Ironically, I used to hate running. Standing at 6'3" tall and weighing 205 pounds, the sport has never come easy to me. I can recall that as a young soldier in my 20's I dreaded early morning runs for those very reasons. At the time I had yet to develop an apt for the mental game required when pushing oneself over long distances to the monotonous beat of simply shooting one foot in front of the other.

Then when I left the Army I declared "I will never, ever, run again...not in a million years!" Yet despite the disdain for pounding the pavement back then, it's funny that a decade on I still get up at the crack of dawn to start my day that way. As the miles racked up, I started to gravitate towards racing and competing as a means of staying fit and exploring new places I visit. But even with that level of commitment, the almighty mecca of running, the full marathon, has somehow managed to elude me.

The item had been on my list since creation, but in the rush to grow up I kept putting it off. Perhaps I was afraid of the pain involved, but always it lingered in the back of my mind. The mantra was to "explore, experience, then push beyond" and so with no route or time restraints to dictate how the expedition would unfold, I knew that at some point, with a little luck, I would cross paths with a full marathon somewhere. And when I did I was gonna have to man up, because the trip's intent wouldn't be met without the accomplishment.

After about an hour of running around town, covering nearly eight miles of well groomed trails, I emerged from the dense woods at Taco feeling energized and pleased to be in time to catch a beautiful sunset. At -10 Fahrenheit the run was colder than I initially thought it would be. But, besides my feet being sore from the thrash of icy paths, and a small spot of frost bite that had developed, the jaunt was otherwise rewarding and set the bar high for the ninety two runs in one hundred days that followed at some of the Continent's most iconic places.

I took a selfie with my phone thinking that in many years I would be able to look back at the portrait and it would speak a thousand forgotten words about the first leg of the journey. As I looked at the picture to study the frame I laughed out loud wondering, "who are these impostors before me?" I was certain that even a professional photographer getting paid to do a shoot on location would be envious of the image I captured that evening.

In it, both Athena, my trusty K9 companion, and I were sporting thick beards of ice on our faces from the condensation of breath, and our profiles burst from the stage of the falling sun, the forest and deep white snowbanks. The overall effect was deceptive and made us appear wild, not domestic, or at the very least like we belonged in that timberland. And although I knew we didn't, the mutating sky above did remind me of a winter I spent in the High Arctic, and so I paused to watch the show above.

For those who don't know, the Arctic is a frozen remote windswept land located at the top of the globe that has an area so massive it completely encircles the North Pole. A harsh place, it's full of peril, blinding sun, endless tundra and sea ice that spans thousands of square miles. In fact it is so far north that trees can't grow in the frost filled earth, even in the dead of summer, because it hasn't thawed in a million years.

Amazingly and despite this unforgiving environment, somehow life thrives there, and at the heart of it, there exists an ancient culture of hunter gatherers called the Inuit. They are inspiring, determined folks with strong family values and an economy still based on trade and goodwill. I fell in love with them and admired the fact that they are truly nomadic folks.

Who among the readers of this book has not at some point spoken of Canada without visualizing an Igloo or romanticizing the image of a strong warrior clad head to toe in fur as he stands motionless for hours waiting for a seal? I make that sound romantic, because to me it is, but the reality is that it's a tough life to eke out up there.

We were on Baffin Island at the time and doing survival training in a realm straight from the pages of National Geographic. The Inuit guide assigned to teaching me was that warrior I speak so highly of, and because my life was in his hands, I showed unconditional obedience to him. I can cite on one occasion after a long haul of hunting and tracking that I had some reservation though. He pointed in the general direction of where he wanted me to set up, telling me in broken English and hand signals to pitch my tent "over there...over there..."

The site was right at water's edge and even before I walked over to take a look I had to conjure some nerves to approach. First off, I had yet to master the skill of shuffling on the slippery bobbing ice that carpeted the Arctic Ocean where we slept. But secondly, and the real reason for my angst, was that polar bears lurk at the edge and I couldn't seem to shake my fear of them. The cuddly looking bruins are the world's largest land predators, and because they are atop the ladder they pay no heed to competitors. I wasn't being irrational either, the warning we got each time we ventured out to take a leak was "be careful....even us Inuit hunters get ambushed here." My guide was a man of few words and didn't appear to be rattled though, instead he smiled and pointed again, "over there...over there..."

"It's gonna be a long sleepless night," I thought as I nodded.

After setting up I needed to recharge with a nap prior to my bear watch shift. Before I did, I grabbed some local sushi from the group to fill my belly. This wasn't the grade of sushi you find in California Rolls though, this was unique Inuit fair. Which means it was fresh Char, pulled straight from the brine, butchered on the ice then consumed raw while still warm. Its sounds unappealing I know, but in a place where the average man pulling a sled over the tundra burns 12,000 calories a day, this fish with its inch thick layer of fat was a real treat.

As I sat there picking small pieces of flesh from bone and observing the frozen expanse, I felt envious it wasn't my home. "What a wonderful spot to have lunch," I pondered aloud, taking in its purity. I prolonged the meal until my fingers went numb and I was forced to head in. Hours later I emerged from my slumber, rested yes but confused that the landscaped had morphed. Struggling to make sense of it all, I noticed my heart was pounding loud enough to drown out rationale. I began to search for a way back before acknowledging the obvious out loud.

"The ice is drifting, I think…"

It was the ocean so of course the tide goes out, but I kicked myself that I didn't realize just how much until now. As my mind continued to race I became intensely aware of the uncomfortable feel of clothes sticking to my skin. My parka was wet with sweat that trickled then pooled at the small of my back, and no doubt I was experiencing a bit of shock. Inhaling deeply to counter its effect, slowly I began to calm and in doing so went through the various stages of grief from denial to acceptance. But because time wasn't a luxury, the process took only minutes versus the days, months or the years it should have. With it though, my heart rate began to moderate and made way to relevant questions like, "am I really drifting out here? Weird though it seems that I might be doing that?" and "Well, this is lovely...so where is my guide at?"

Really I wasn't alone, I could see others from the group on different packs of ice nearby, but in the moment it felt like they were a thousand miles away. As I took a knee to allow my body to catch up, I could hear laughter coming in from all directions and wondered if it were meant for me. "Of course it is!" I mused, realizing I had been stumbling around like an idiot with less grace than Chris Farley's character "The Motivational Speaker." I was relieved that no one else was concerned, but couldn't help but feel a tad foolish in remembering we were taught that when the tide goes out, it does eventually come back in. I was in no danger of being anything but a laughing stock in front of friends.

The experience up there was a once in a lifetime one, but what I took from the Arctic that still persists today is my passion for watching sunsets. I remember the joy I had at the end of each day up there when I could walk to the water's edge to capture one. It became ritualistic, and I would lower my body down, prop my back against anything hard, then fling my legs over the fringe to allow them to dangle freely above the blackness of a ten thousand foot abyss.

Then after gazing down to try and spot a white monster charging from the depths, I would relax to the sound of water lapping against my mukluks and look to the horizon for my gift.

There in the distance was the same timeless ballet I was observing in Whitecourt now. The pastel colours of dusk were upon us and being cast through the thin layers of clouds way up in the atmosphere, where they seamlessly transitioned into the dance of the Aurora Borealis. You don't have to be immune to the cold to see there is beauty in the Northern Lights.

I spent the next couple of days staying with my Aunt Linda and Uncle Rick and enjoyed their company very much. They are both business people enjoying plenty of success and so we talked a lot about the economy and really just caught up. Before leaving I packed some gifts the family had wrapped and asked me to drop off to my sister when I got to San Francisco. I then thanked them for everything with a big hug, and headed off for one last stop in Fox Creek before I left the Province.

* * *

My Uncles John and James are both cowboys like my Grandfather was, and just like with Gramps I hung off their every word. The two made for easy idols, and I recall my hobby back then was to mimic the walk, talk and mannerisms of western culture they emitted. Rodeo stars once upon a time, I'd brood for hours like a pest to absorb their unique banter as they and their friends honed riding skills on a "mock-bull barrel" out in the yard. That barrel was actually just an oil drum strung from trees that tumbled and spun as each took turns trying to buck the other one. I was their biggest fan, and so I was looking forward to catching up now.

The drive to Fox Creek was a short one and I arrived around supper time to be greeted by Uncle John and Aunt Nadine. Both are extremely kind, hardworking and down to earth people who leave a lasting impression on anyone they meet. John is a hunter who has the ability to charm with tales that have always drawn me in, and it wasn't long before he had me captured again. Over drinks he told stories of close encounters with bears, wolves and cougars in the forest. They seemed tall, but I knew they weren't, instead they were just foreign to me now, and as I listened I felt the same comfort I had as a youngster return. I realized that the affinity I had for him and Uncle James, who wasn't there, was justified and that although it may have been lost with age, it was replaced now with high regard.

The next morning I was awoken to a voice in the dark. "Aaron... get up...we're going hunting." I'm not much of a hunter, so confused, still half asleep, and probably drunk, I responded reluctantly, "Nope, can't go...not licensed to hunt." It was a diplomatic way to say, "Leave me alone, I'm hurtin'," but the argument didn't pan out well. Bellows began to ring out and I was sure somewhere in the house people were mocking my agony. Rolling over to defend myself I could see the blurry figure of a man in the doorway, his voice spoke again but more clearly this time, "Just tag along if ya can..." he said with a hint of sarcasm.

Awake now I started to smile and flashed to the night before where in a half cut state, we had discussed the early morning outing. I recalled feeling quite confident then, and dared anyone in the room to wake before I did if they thought they could. A tall order given it was spurred by liquid courage that tilted the odds in their favour with each rum I drank.

Heckling continued to pour in from down the hall and the light in my room switched on. The act momentarily blinded but revealed the tormentor. There with a rifle in hand and a grin on his face was Uncle John. "You slept in, army boy," he said smugly, then turned to walk away while uttering in twang, "Get up...let's go." Bested, I stumbled out to give credit and to beg for a coffee. We were in the bush within an hour.

No deer came home that night, but with a sack of wild chicken to feast on, the jaunt was well worth it and a lot of fun. I left early the following morning and backtracked South by following the Cowboy Trail along the Foothills, then went West via the Bow Valley, through the Rangeland and into Canmore, Banff, then entered British Columbia. The Alberta Leg of the trip, while not quite like the adventure blitz of stretches to come, was filled with family, friends, and familiarity, and was the perfect segue to compel me on. Coastal bound, I was finally on my way into the unknown.

Leg 3
Surfs Up In The Wilds Of British Columbia

"Believe you can and you're halfway there."
- Theodore Roosevelt

British Columbia on Canada's West Coast claims to be *"The Best Place on Earth."* Certainly a message not lost to curious tourists hoping to capture some of her charm, and clearly one visible as you cross the border with Alberta and are greeted by provincial licence plates that proclaim "Beautiful British Columbia."

Often referred to as California North, the region is known for its natural splendor and cultural diversity with influences not only inherited from across the country, but around the world. Its enterprising citizens are a unique combination of immigrants, native peoples, farmers, hipsters and pot smoking hippies who all share a grand landscape of deserts, lakes, mountains, forests and pristine coastal waters.

It was that abundance I was chasing as I headed towards Vancouver via the Trans-Canada Highway that day. Normally a beautiful scenic drive, the route seemed more treacherous than therapeutic. Falling snow, poor visibility and slick conditions through the mountains made for a long trip as big rigs with heavy loads immune to ice and snow flew by at an alarming rate, leaving in their wake glaring lights, blinding slush and a sense of dread as I pushed deeper into the heart of the province.

I had expected such trials on the trip, but the stress of that one made me question my timing. *"Perhaps in hindsight I could've picked a better day to tackle the Pass!"* I thought with contempt referring to Rogers Pass, one of Canada's most notorious stretches. The pass itself is located in Glacier National Park; a narrow, undivided section of road that sits atop an imposing 4,400 foot summit. People come from all over the world to experience that drive and are rarely disappointed by the unmatched vistas that appear around each bend and go on forever.

We're not the only ones who see value in a conveniently placed asphalt trail carved through the wilderness though. Often tourists get more than they bargain for when they cross paths with elk or bighorn on the side of the shoulder and swerve to avoid them, but end up instead in oncoming traffic. Fatal accidents are common and the notion of being in one didn't escape my thoughts as I continued to navigate the complex system of bridges, tunnels and avalanche slide zones I encountered along the way.

Once through the pass, but not even close to where I thought I'd be by dark, I decided to pull into the next rest area to get some sleep for the night. Exhausted, I crawled into the back of the Taco and remember feeling a sense of relief at having the stressful day behind me. Shutting off my flashlight, I was asleep before my head hit the pillow.

The next day I awoke rested, recouped and feeling the kind of satisfaction one gets from a rare deep sleep where you wake in the same position you lied down in. Rolling out of the bag I stretched then reached over to open the tailgate hatch and to my surprise was slapped by an awe inspiring sight, twelve inches of fresh snow had fallen through the night. The effect, combined with the bold sunlight, seemed to magnify and define every crevice of the slopes.

I was parked just outside of Revelstoke and although I wanted to get moving, the small historic alpine hamlet nestled along the banks of the Columbia River was seducing. Literally the town's reputation as a winter wonderland precedes it and although encircled by towering white capped peaks, likely the first thing you'll notice, as I did, is the giant statues of grizzlies that stand sentry, then lead folks in.

After following their directions downtown, I parked and soon found myself strolling the streets as a lost and carefree tourist with a coffee in one hand and a crumpled up map in the other. As I walked I met a lot of friendly people, some of them shopkeepers, others longtime residents and still others who were traveling ski bums from all over Britain's Commonwealth of Nations.

Moving along I stumbled by accident onto Mackenzie Avenue where I found myself snapping one picture after the next of the unique architecture seen in the old buildings and quaint businesses that line and reaffirm the town's identity. I could feel my feet becoming wet by snow that hadn't been cleared yet, and as I hopped down an abandoned street near the old railroad, my mind began to conjure.

Prompted by the encounter with the ski bums, I imagined a transient from a different time, and began to follow in the footsteps of a homeless man whose ghost I reckoned was a drifter from the Great Depression. His days, although condemned to loneliness and riding the rails in search of room and board, were for some reason, enticing to me. Captivated, I smiled thinking our journeys seemed the same, but I knew they weren't, both nomadic yes, but the parallels ended with that. My trek of the continent was born of pleasure and his of necessity. Although removed from his struggle, I couldn't help but be envious of the freedom he had. The town had that feel.

I had been to Revelstoke many times, but had never taken the time to explore it like I did that morning. Always in a hurry I would pass this gem by to get somewhere else. As I fueled up and prepared to leave town, I thanked the Universe for the snowstorm that forced me off the road, and realized that moving forward I could explore any burg I choose in a similar fashion.

Already I was beginning to appreciate all the small things I hoped the road would offer.

On the move again, the weather was improving as dramatically as the changing scenery. Where just that morning snow had blanketed the elevations, by noon I was at Shuswap Lake where there was none, and by one o'clock I was passing through the arid oasis of the Okanagan Valley's Wine District. Now cruising the Coquihalla Highway, it was warm rain that invited, and with my windows rolled down I closed in on the Sunshine Coast. It felt like I was a million miles from home.

After a quick stop in Hope, the small town where "Rambo First Blood" was filmed, I pulled onto a gravel road then followed it for about a mile looking for a place to let Athena stretch her legs. As we drove down the road I noticed she had her head out the window and curiously smelled the air.

"Strange...she never does that," I thought, but didn't really care. Once parked, I walked around to open her door to let her out. She bolted past me and in the blink of an eye disappeared into the dense undergrowth.

Concerned for her safety I tore into the tree line after her but didn't make it twenty feet before stumbling over a log that had been concealed by high ferns on the forest floor. The momentum caught me off guard and propelled me forward head first down a small bank where I landed hard at the bottom into a dried up section of a larger river bed. Confused for a moment, I laid battered and bruised, but okay as I scanned the ground wondering where she could be.

Athena is a robust dog, shy at times yes, but never hesitant when it comes to opportunities of flight. Countless times I've seen her at the sight of a lake dock, run full throttle to its end, then with her best Superman, belly flop into the chop. It's a funny sight to see but sometimes, like now, it's nerve racking as well.

Off in the distance, about a hundred yards to my front, I spotted activity in the water and to my relief could see her frolicking about. Relieved I walked towards her, thankful she was okay but wonderstruck as to why my normally reserved dog was acting so oddly. Nearing I could smell the unmistakable scent of death being carried on the air and at first thought it to be a dead deer that lured her here. Soon though I realized the temptation was much more persuasive than even that, because scattered all along the water's edge were the remains of dead fish.

Many of them were clearly rotten and had died some time ago but others looked fresh and appeared to have been plucked within the hour. Their bodies were still intact but unnervingly their heads were missing and I could vaguely remember a lesson from grade school on the cause of that. "Wolves," I said with dismay in my voice as I studied the tree line for signs of the pack. I knew we had to get back to the truck and fast.

Approaching Athena, I could make out strange activity in the river and shadows began to play havoc with my eyes as objects darted to the surface, then back into the depths again. I had rubber boots on and drawn to the commotion I edged slowly into it. Once there I could see a drop, and that the darting shapes were in fact big colourful fish in numbers so great their mass seemed like one. They appeared to be competing for space and in the process brought additional life to the river with the chaotic sounds of shifting boulders being rolled under them.

Some were as wide as my thigh and in places packed like sardines, and in the melee they were brushing up against me. The confines shared with them was a welcome bond that both impressed my senses and acted to test my balance in the swifts, and at least once I stumbled back from their thrust. The impact of which felt as wild as the confines of British Columbia, and worked to send primal chills up my spine with pulses of energy. Standing there I realized we had stumbled inadvertently onto one of Mother Nature's greatest gifts, a West Coast salmon run.

Poetically these fish were born here, left to circle the globe, and now have returned home to spawn in the very same tributaries they were from. Drifting towards me, just at the surface, was a large bright red male with a pronounced hook nose and a broken fin that made him stand out. I could see his eyes held genuine knowledge, but they appeared static and glazed and it was evident that like the others he too was dying. Reaching down to cradle his fat belly, I became intrigued by the scars on his back and gently pulled him from the water.

In the back of my mind there were wolves in the area so I didn't want to stay long, but took a moment to run my hands over his scales and healed scars. I wondered where each one came from and what tales they could tell of his journey before we met by chance on the river that day. He was labouring to breathe so I didn't ask, but thanked him for the brief chat and placed him back where the current soon took him away.

Refocusing my attention I grabbed Athena gently by her collar to pull her over, instantly her ears went back and her look of innocence turned to guilt. Once ashore and dripping wet, I solemnly knelt before her, looked her in the eyes and said, "Don't run away again sweetheart!" Her impulsiveness had led us to the water and to the unique experience that awaited us there, which was a reward for sure, but I was determined not let her out of my sight again. As I drove the short distance to Vancouver, I knew I would have to be more vigilant moving forward, and I'm happy to say that for the rest of the trip she never bolted quite like that again.

* * *

Vancouver is an Olympic city, one of several I would encounter on the trip actually, and like the others the title comes with certain connotations. Clean, friendly, rich and with a mild climate, the community usually ranks among the top five cities in the world for livability. It's an electrifying place, but it was the West Coast surf that called my name, not the big city lights. So after a couple of days of runs along the Seawall in Stanley Park and pub crawls through Gastown, I was soon on my way and bound for Tofino via a ferry that promised surfing dreams.

The crossing to Vancouver Island was nice, but arriving after sunset to a place known to sport a lot of wildlife, I decided not to push the envelope by driving at night and instead pulled over to set up camp. The spot I found was only about five hundred yards down a dirt track that was off the highway, but dark enough to evoke claustrophobia. With the pressure of Athena's back against mine to keep me warm, and the sound of tapping rain to pull me into REM, like in Revelstoke, I was asleep in no time.

I woke the next morning feeling overheated and with the windows painted by a thick layer of condensation. Throwing, then kicking my sleeping bag off for relief I realized Athena was panting too, a rare sight given that she hardly ever shows signs of stress even on hot days. "It was the rain last night!" I rationed with a grin, acknowledging its calming effect and the culprit for me sleeping in.

Because I pulled in at night it wasn't until now that I was privy to the bounty of the island. Opening the topper to let the heat out I was greeted by a magical landscape adorned by towering old growth trees, giant ferns and layers and layers of moss all being smothered by the intoxicating smell of cedar soaked earth. Breathing deeply to absorb it, I smiled that for the first time ever I was in the Pacific rainforest.

I lowered the tailgate, then jumped down, and was numbed for a moment by the clutter of holistic sounds. I could hear tiny beasts scampering through the undergrowth and waterfalls crashing about. The activity added more mystery to a place I knew little of and reminded me of my trips to Hawaii earlier in the year. Although I loved that place, with this at my fingertips, I wondered why in life I always seem to stray so far when often what I'm looking for is right here. I had woken with Tofino and surfing on my mind, but after seeing the forest for the first time I knew I had to explore by putting a few miles on.

After changing into my running gear, we headed down the dirt track away from the highway and deeper into the unknown. Within a few minutes the road narrowed almost completely until it was little more than a slippery, muddy game trail that weaved through the trees clumsily. Because it forked in different directions, all of them looking the same, I felt disorientated and began to snap twigs and draw arrows in the wet ground to help guide me back, sensing as I did that we were being watched.

I continued down the trail for about three miles before turning around to head back to the truck and as I neared the end of the route noticed something alarming in the mud--fresh bear tracks. They hadn't been disturbed by the previous night's rains, nor did I notice them on the way out. As I crouched to examine them closer I could see that some of the tracks overlapped my own and so deduced they couldn't have been more than an hour old.

Growing up in Alberta and being one that likes to escape to the backcountry whenever I can, I've crossed paths with these bullies a few times. But never in areas where I didn't feel completely at home, and usually while having the benefit of seeing them first. Pausing for a moment to assess, I slowly rose from my knees to a higher vantage where I attempted to peer into the dense woods in search that a dark figure might lurk.

The effort was pointless though and only added to the anxiety. With my eyes strained my mind began to race and soon played tricks on my consciousness. The unrelenting shadows, sounds and movements of the forest hinted ubiquitously in the predators favour and, paranoid, I was forced to surrender to the idea that we were being stalked on the trail.

Moving again I let Athena lead the way so I could see her and quickened my pace to keep up. I knew the truck wasn't far and that I could sustain the increased effort for at least a few minutes or until my thoughts were drowned out by the burning in my legs. Coming around the last bend I could see the back end of Taco poking out in the distance. The sight gave me a renewed vigor and I dug deep to sprint the last few hundred yards to safety. Once there, a feeling of relief came over me as I fought the urge to collapse to me knees. Leaning over, I wheezed to Athena, "We made it!" then "Was it all in my head? You woulda told me if there was danger out there…right?"

She stared back at me blankly, groaned then canted her head sideways to suggest aloofness. "Or not," I admitted, laughing out loud as I remembered back to a time I watched a mouse eat her lunch from right under her nose as she slept a summer afternoon away. Athena is a lot of wonderful things; comforting, quiet, kind, patient and sweet to name a few, but guard dog she is not, and likely never will be.

After quickly getting changed, I opened the door to let a now very soiled dog up onto the leather of the front passenger seat, then walked around to the driver's side. Firing up the engine I did a three point turn on the narrow stretch to maneuver us back towards the highway again. While finishing the turn I punched the gas to fishtail out of the mud, and as we began to spin around, I slammed on the brakes to bring us to a screeching halt. There less than twenty yards to our front, and in the middle of the road with eyes piercing straight through the cab, stood a black bear.

"I knew it!" I yelled out as I slapped the dash to say I told you so to Athena who now noticed the bear too and was growling through her open window. "Ya right...now you're tough," I said laughing as I patted her on the back. The bear was small and likely a lone female. After observing for a moment I pulled forward slowly to try and capture a close-up, but as I did she bolted back into the shadows from which she came. I shook my head impressed again by the wildness of British Columbia.

The drive across the Island seemed long but reflective and although it's only about fifty miles wide, it took me more than four hours to navigate the twists and turns of the Pacific Rim Highway to get to Tofino. Located at the extreme western point of Clayoquot Sound and at the tip of the Esowista Peninsula, Tofino is a small ocean-side town of some two thousand that's famous for year around surfing, whale migrations and winter storm watching. Surfers, with their guarded culture and connection to sea, tend to flock to the area for its isolation, consistent swells and familiar country feel.

By the time I parked it was midafternoon. I was a bit hungry but anxious to explore so I crossed the street and walked a short distance to the Tuff Beans Coffee House on 4th street to grab a quick cup of joe and a sandwich to hold me over for a couple of hours. The cafe was small, quaint and colourfully decorated with a girl working behind the counter who was a young surfer type with thick dreadlocks and who spoke in a calming voice. As I neared she pointed at the chalkboard menu above her head and asked me what I'd have. Because it was chilly outside I said, "Anything hot and with some lead."

"Dark roast it is," she said.

As she grabbed a cup and started to fill it I ordered a sandwich as well and wondered why she was smiling so much. Returning the gesture I asked, "What's so funny?"

"You're not from around here are you?" I knew we had never met before but remember thinking, *"is it that obvious?"*

I answered, "Nope...what gave it away?"

She stopped what she was doing, looked up at me and said with a hint of sarcasm, "You have a weird accent, you know..." then beamed and nodded in the direction of Taco.

I veered back to see the tailgate was visible and said, "Ahhh, right...Alberta plates...dead giveaway."

With the ice now broken we engaged in small talk for about ten minutes as I waited for my sandwich. She told me her name was Jen and she was a single mom, and I told her about my trip, my surfing ambitions and my goal of exploring the town. The conversation was good and came easy and as I got ready to leave she suggested a couple of pubs for me to check out later. I grabbed my food, thanked her, then walked out the door to explore the rest of the day.

The next morning I was up at the crack of dawn. By 5:30 I had a coffee down range and was out the door headed in search of the rainforest hiking trail located in the Pacific Rim National Park. Once there I did a short run to start the day and afterwards headed back into town feeling alert, hungry, and determined to meet a willing surf instructor.

I stopped by Tuffs again for breakfast and was glad to see Jen was working the counter. After telling her my goal for the day she pointed me down the street. "Try Westside Surf, it's owned by Sepp and he's a great instructor to any level, plus they can usually get you in pretty quick this time of year." I thanked her again, and when my coffee came headed out the door.

The shop was right where she said it was, and in fact being a small log building with huge glass accents attuned in surf boards and a neon sign that read "Westside Surf," it was hard to miss. On entering I approached the register and a man who was kneeling down stood up to greet me by asking if I needed anything.

"Sepp around?" I replied as I forced my hand to shake his.

"You got him," he said coolly with a nod while extending his grip. "Oh perfect! Jenny from the block sent me over," I said, pointing across to Tuffs.

He laughed. "Right, of course...surf lessons I suppose?"

After introductions, I explained my trip and again my surfing aspirations. As soon as I was done he turned to another staff member and said, "That sounds like an awesome time. Let's get Aaron out before the weather turns...book him with Greg for this afternoon." Thinking that the lesson would be the following day and taken aback by his accommodating way I must have paused because he looked at me square and said, "Are you okay with that?"

Not wanting to appear nervous I responded as boldly as I could, "Um, yeah, of course... that'll work!"

The truth is that I've always had a fear of open water, coming from the arid regions of southern Alberta where water is scarce and often restricted the thought of being in the endless ocean was terrifying to me. Still the weight of the experience appealed more than the fear and after dropping Athena off at the hotel room I rented so she could wait the day out in comfort, I headed back to get sized up for a wetsuit, a surfboard and to meet my mentor, a tall Hawaiian man in town for the winter named Gregory.

He greeted me at the door as I arrived and I was glad he did. With butterflies through the roof as the adventure neared it was refreshing to meet my veteran teacher. Although a man of few words, he possessed a rare demeanor that exudes fortitude and in turn commands a Zen-like quality. Even though we just met, already I trusted him and could see he would be the ideal ambassador to introduce me to the sea.

After some instruction in the shop the two us grabbed our boards and gear, and headed out back to load up for the ten minute drive to Long Beach via the Coastal Highway in his charming little Toyota pickup. Beat up by decades of better days, the vehicle made no attempt to hide its labouring ways as it rattled over bumps and past ruts along the way. Glancing in the passenger side mirror I smiled as I caught sight of the surf boards bouncing on the tailgate behind us. The scene in the mirror looked iconic and it could have been from any one of a hundred Hollywood surf films I'd seen growing up, but this one was written to star me.

We pulled up to the beach, grabbed our boards and walked the couple hundred yards to the water's edge where I could see in front of us a number of surf gods already bobbing away, it felt intimidating. Before entering we spent a little while rehearsing drills again and going through proper posture. Greg tested me on what to do in rip tides, and how to spot bubbles when tumbling. We did that until he was sure I was comfortable with the task at hand. Well into the practice I started to catch on and noticed his attention beginning to drift to the horizon, I could tell he was studying the conditions. After a few minutes he looked back to me, grinned, and with satisfaction said calmly, "Let's go...surfs up, bro."

That was my cue, I jumped to my feet as a wave of adrenaline shot through my body and, *"If not now, then when, Aaron?"* permeated in my head. Knowing there was no going back, I gritted my teeth, manned up, and willingly followed Greg in. Venturing out past the sand bars to open water we waded until, to my horror, the ocean floor was no more. The water turned dark and the sandy bottom disappeared. Once in location I pulled myself onto the board to straddle its girth and felt vulnerable at first wondering what creatures might lie beneath. I was aware that because of my phobia I shouldn't be thinking such things, but after years of avoidance it was actually liberating.

With Greg's direction I began with some success to chase waves and tried over and over again to stand and harness their power. Lacking the benefit of balance, each time I would be tossed headfirst into the frothy water, where with stinging eyes and a mouthful of brine I struggled to surface. After hours on the water and still not making progress I was becoming tired, discouraged and needed a breather, so went to the shore to think things over. Greg, sensing my resolve was waning, followed suit to offer words of encouragement.

"A few hours ago you were afraid of the sea...now you're spending all your time upside down in it," he said to tease me.

We both laughed, but the comment did spur the belief that I was already halfway there. Refocused, I watched the ocean swells in the far distance rise and fall before rolling and crashing into the shore. *"It's one wave for the rest of your life,"* I told myself to buy nerves as I swam out to take my place among the tide. The "rest of your life card" is one I've pulled many times when pursuing goals. It's a reminder that nothing is out of reach, and that if I don't cross the bullet off the list now, I might regret never crossing it off.

In position again, and with my attention wholly on those incoming swells growing with each passing yard, I turned my back to them to point my body and board towards the beach. As I did I could feel Greg push me away while yelling, "Paddle!

Paddle! Paddle!" I did and like a freight train I began to gain momentum. In the corner of my eye I could see the water beginning to rise higher and higher until finally it began to roll into the familiar white surf of the pipeline that had already bested me over and over. No question it was time to stand.

Mindful that I didn't want to start the process all over again by thundering in, I used slow deliberate movements to shift my weight and placed my arms squarely under my shoulders as I was taught. Then with one quick violent motion, I forced my body up from the board by sliding my knees forward until they were just below my chest in a crouching position. Frozen there I consolidated for a few seconds, then with wobbly legs rose like the mighty Phoenix.

Upright now, but barely, my confidence soared and time seemed to slow to a crawl as if to allow me an epoch to absorb the significance of it all. Where throughout the day thoughts seemed chaotic, crowded and rushed in my attempt to sequence the drills correct, now they felt logical, clear and rudimentary. For the first time I could feel the effects of wind and mist slamming against my chest and with a bird's eye view, I watched joyfully as water passed under the nose of my board to carve a wake of white foam beneath me. I was now walking on water.

With arms fully extended out past my sides for balance I mimicked the image of an untethered 1920's iron worker, and embraced the atonement earned after hours of failure. Laughing out loud at myself I thought, *"Oh my God! I'm finally SURFING!"* I shook my fist at Greg to catch his attention and as I did stumbled forward, tripped, then bailed into the cold water to end one of the most gratifying experiences I've ever had.

Ok, so not the epic barrel rides of Patrick Swayze in his movie "Point Break," but I did stand, albeit awkwardly, and for about six seconds too, so I was happy. We went out a few more times and I had more success with each. Greg suggested afterwards that if I was to spend another day with him on the water I would improve dramatically. Although the prospect did appeal, after some consideration I had to decline the offer. The trip was nomadic by design and the intent to "explore, experience, then push beyond" was meant to come in bits, not bites. I wanted to try new things, not master them, and so with surfing checked off my list, it was time to keep moving on.

As we drove back to town there was little conversation between us. After six hours on the water I was tired, sore, and impressed by the athleticism of surfers. I got back to the hotel, showered and changed, then headed to the Shelter Pub for dinner. There I met some friendly locals who were surfers like me…but the real kind, and we talked shop over a few too many. The next morning I was off again, this time to explore the cities and spaces of America. Things were starting to unfold in a random way.

Leg 4
The Oregon Coast And Border Delays

"Try to be a rainbow in someone's cloud."
- Maya Angelou

Patiently waiting and watching through the window of the American detention office at the Peace Arch Border Crossing just south of Vancouver, I wondered to myself, *"What is it about me that always garners the attention of Customs Officers?"* A good question considering it was noon, and Taco was already three hours into the indignity of a full body cavity search that saw our possessions scattered across the sidewalk. With morose I couldn't help but ask Athena, "Do you think I could've been a little more cordial with the guard today?"

Let me turn back the clock about four hours or so to earlier that morning. After idling for some time in the long lines of traffic trying to enter the United States, I pulled forward relieved to finally be arriving at the guard house. Once there and eager to keep moving, I decided to employ some charm with the female officer inside by unleashing a brand of small talk I was positive would both flatter and see me on my way. Turns out though, I ain't got enough game to be a "Playa."

After coming to a complete stop I rolled down my window, looked at her, and while handing over my passport said the first unmindful thing, "Hey, good mornin'...you look really tired." Picture my emphasis on the words "really" and "tired." I was alluding to fact that it was a busy day, and for her probably a long one too, so the comment was supposed to do nothing but show gratitude. In hindsight however, I could've picked my words a little wiser.

The tone in my voice combined with a poorly timed smirk on delivery did no justice to my sincere intent, and instead I'm sure I came across as an arrogant jerk. Suspecting that was the case, I struggled to recant, but with each attempt did nothing but dig myself in deeper. With her hands now resting squarely on her hips, and her eyes sending daggers through me, I was certain of just one thing, I had offended this nice woman.

I respect officers of the law and value their work immensely, understanding like most that it's no easy task to protect the lives of others while offering the unlimited liability of their own. Highly trained and qualified professionals, they are proud products of a unique environment where applying their trade can be both a burden and a reward. Generally speaking, it's accepted that police have high standards of integrity and so strive to enforce the rule of the law in a fair, just way. Still, even cops have bad days, and knowing that she could deny me entrance for any reason at all, the fact was weighing heavily.

Inhaling deeply through my nose to conceal angst, I started to relax and could rationally reflect back on the events that had taken place. I discerned that what had happened was nothing more than a misunderstanding, and reckoned that likely she had interpreted my choice of the word "tired" for a lesser known synonym from the "Feminine Dictionary" to mean instead, "disgusting."

This wasn't the first, nor would it be the last time I've failed at attempts of "Womanese." But aware the blunder was gonna cost me my day, I looked straight to my front, sighed in frustration, then uttered under my breath, "That DID NOT come out right."

As she turned away, my conscience drifted and I began to think back to lessons my mom had endeavored to teach her boys on all things girls and good etiquette. She had a lot of wonderful attributes to offer; intelligence, empathy and dogged determination to name a few, but strong paternal figure she was not, and she'd be the first to admit it takes a man to teach a boy how to act. Being a single mother though she had many hats to wear, including that of an absent dad, and I can recall a time when on the eve of my first date, she sat me down and to my dismay, began to talk about what else, but "the birds and the bees."

As a reader you know that by now I embrace the uncomfortable. It's a handy trait to have and one that usually works to my advantage. But even for me the silent moments of that conversation were unbearable and as the minutes passed like hours, I recall wishing for nothing more than to end the discussion I was having about sex with my mom.

For about the last half hour or so I had been sitting quietly on the couch, where by scarcely moving I had hoped to expedite the process. Typical of her blue collar roots, and to calm her own nerves I think, she had in one hand a smoldering cigarette, while in the other a glass of red wine poured past code so it splashed freely in concert with her passionate arm movements and shimmering "Weird Al" hair.

Although the theatrics seemed dramatic, I could see she was trying in a roundabout way to say men need to be good mannered and chivalrous with the ladies. Near the end, and seeing that my attention was drifting, she leaned in to give one last tidbit, "Son... with a woman never ask her age... don't mention her weight... and do NOT, under any circumstance... EVER!... tell her she looks haggard."

Decades later, and with the trip of a lifetime stalled at the border and hinged on a woman's scorn, I finally realized the value in those words. Smiling I could sense that somewhere, somehow, Mom was giving me the same "I told you so" look she had mastered so well. Feeling the sting of consciousness, I bit back, "She's right, but in my defence, what I still don't know about women could fill a book..."

Turning my attention back to the officer I attempted humour, despite knowing it could seal the deal, and said, "Soooooo... where's the secondary inspection at around here?"

The query triggered a slight, albeit malevolent twinkle, and in a friendly tone she responded, "Straight ahead sir. Just follow that officer, he'll take care of you."

"Lovely," I fired back to insist I was certain he would. We both shared a discrete chuckle over the gaff, and after gathering my passport, I slowly pulled ahead. *So much for flattery at the border,*" I thought.

It's fair to say that I've never fully understood women. Like so many lessons in my life, it went untaught without the presence of a male role model. The absence led also to confusion on what the measure of a man is, and it didn't seem fair that everybody else I knew had two parents. Feeling like an outcast I became resentful, even mad at the world for it. I suppose all kids perceive that same awareness of not fitting in as they begin to find themselves through the trumpet of low self-esteem. My mom, seeing that perhaps I was struggling with confidence at times would always spot an opportunity to be my rainbow and would say, "What is normal, Aaron? Just be proud of who you are, kiddo."

My father was hardly the same nurturing soul. An alcoholic, a batterer and an abusive man in every sense, he took no guise in raising us and was in fact absent my whole life from about the age of five and up. Not that I had any choice in the matter, he didn't really take an interest in exercising his right to visits. Despite that, the distance only worked to fuel my fascination, and for years I had him on a podium. Anyone who has ever lost touch with a parent at a young age will tell you that there is an overwhelming sense of loss amplified by tear-filled nights and a deep longing to reunite, even if wrongs were committed.

Convinced that he too felt the same way, I remember running home from school to rifle through the mail, hoping that at some point a letter from him would come. The act was compulsive though, and by definition crazy as it only ever yielded the same result of nothing. It wasn't long before my siblings gave up on him, but for some reason I continued to hold on.

I used to lash out at my mom for leaving him, not understanding until I was grown that her actions were out of desperation. You see, for nearly a decade she had endured emotional abuse, neglect, broken bones and broken promises. So it was a logical move to leave and one she had attempted many times. Each ended with her either being caught at the door, or by her returning in the middle of the night, lost, alone and bound by the same heinous dependence that forced her out. Then, from him, came the usual promises of "never again." But everyone knows the cycle of abuse doesn't end with an apology, instead it worsens, and like an elephant in the room, the crime becomes tolerated again and again.

I can vividly remember the last time he attacked her. By all accounts it was an average 1983 Sunday afternoon. I remember the year because it would be my last in a two parent home, and the day because there was an aura of excitement in knowing Dad was home. As the house filled with the comforting smells of Mom's fresh bread rising in the oven, Dad joyfully watched over us kids while we entertained ourselves in the rumpus room of that tiny 1,100 sq. ft. bungalow, run down and nestled between the train tracks and industrial corridor of Southeast Calgary.

Somebody had pulled the red drapes of the front bay window closed to reduce the sun's glare. The effect on the room was arcane and laboured to paint the walls a shade of blood red that featured swaths of the occasional beam of light still able to penetrate the expanse, revealing airborne dust from children at play. The Old Man had been drinking casually most of the morning, but was attentive, and I remember him turning the dial of the furniture style TV to muted flickering images of Sesame Street as tunes belted from a vinyl record.

Outside rang the abiding chime of air bursts whistling from a Canadian Pacific locomotive as the engineer touted the train's horn three times to warn of the approach of his Iron Horse. Usually the buzz would stir excitement and send me scurrying out the screen door, where I would sprint across the lawn, through the corn garden, then past the shed to finally arrive at the wooden fence, out of breath but full of zest. Once there I'd keep pace as best I could with frenetic waves to distract the conductor until he noticed me mimicking his horn pulling act. The goal was to entice one more lengthy blast, and to his credit, he usually smiled, then gave back.

It seemed the image of domestic bliss, but I knew it wasn't and even at that young age I had observed the annulling effects alcohol had on my Dad. With its use he became a charlatan whose normally agreeable character could morph from Dr. Jekyll to Mr. Hyde. One minute he would be approachable, affectionate and outgoing, the next cold, distant, angry and impervious to emotion.

True to form, and with a sudden change in his demeanor triggered by the train, he stood abruptly, walked to the turntable, and like a teacher raking nails across a chalkboard, dragged the needle over the records surface. Frozen silent by it, I could almost taste the tension manifest as the look on his face transitioned. Then with diverted attention from us, he steered his eyes down the hall and towards the kitchen where Mom was going about her business.

Grabbing his drink he began to walk to the other side of the house, yelling harassing taunts as he did, "Judy..Jude! JUDY! Where are you?" Looking back I see the signs of his aggression were obvious, but because he was Dad, like any small child, I followed. A verbal quarrel erupted and knowing I wasn't allowed to see them fight, instinctively I slowed my pace to a crawl and approached by sliding one foot at a time. Edging closer, my muscles grew tense and I could feel the souls of my feet tingle with the vibrations as he stomped about like a buffoon. Natasha, then my infant sister, and who I was on route to San Francisco to see now, was in her high chair and joined the melee with screams of her own.

Arriving to the entrance, I stood quietly beside the stove for a short period then with a daring surge of courage peaked around the range to meddle secretly in their affairs. From that vantage I could see the orange laminate dinner table with silver trim flipped on its side, and the loaves of bread cooling there scattered everywhere. Pops was standing provocatively only inches from her, and like a bully at school who steals dignity by crowding space, he attempted to intimidate by spitting threats in her face. Wise to his game, she remained rational, measured, and calm by offering only well-rehearsed responses that had worked cleverly before. The attempt was fruitless however and only incensed him more, and knowing what was next she simply closed her eyes, I think to pray. For a tormentor though, that's a green light to go and with a closed fist he swung as hard as he could to land a vicious hook.

The impact was as devastating as it was predictable. Instantly her body went limp, barreled forward then smashed head first into the ground to create the most grotesque thumping sound I've ever heard. In a rage he wasn't done yet, and despite the fact that she was nearly unconscious, he continued the barrage by kicking and punching her body as he dragged her around the kitchen.

His actions were appalling and an absolute betrayal of the person he swore before God to protect until death. As a witness I felt helpless, bewildered and afraid, and with tear clogged eyes, I recall stumbling towards them. It's a sight that even he knew no child should ever see, and my approach must have ignited a sliver of remorse, because he did stop beating her. Standing motionless now, his temper turned to shame. And when she came to he simply walked away as if nothing had happened in the first place. Like every other fight, the contest was over as fast as it begun.

We left for good after that, but some years later, at his request, my brother Nick and I did meet him, in an empty parking lot. I felt strange in not recognizing his face anymore. He was short, round, bald and no longer seemed tall enough to touch the stars. I knew he was reaching out, but it was too late, we were grown now and no longer needed him, and I could see the pain in his eyes as he realized it. The once mighty lion had been reduced to a lamb that now lived a lonely existence full of regret. I forgive him for the abandonment and lessons untaught, but my loyalty to Mom was just too strong to let him back in. She broke the cycle of abuse, and not a day goes by that I don't thank her for that gift, but still it's clear, the legacy of growing up without a father will always haunt my ability to charm the fairer sex.

There were no issues raised on the search, and once on the move again I felt beside myself at beginning the American leg of the trip. I blew through Seattle and the Cascade Mountains of Washington State before getting to Oregon where I gave my head a shake. There I had to remind myself to slow down, explore and seek those places I set out to conquer. Once I did that, things just seemed to come my way, and the discoveries never stopped.

It was the case when after crossing the mighty Columbia I stumbled by chance into the "Goon Docks" of Astoria. Recognizing the landscape instantly I reckoned the town had changed little from how I remembered it, and I could see why the place left such an impression. I was referring to the fact that it was the backdrop for the iconic 1980's motion picture, "The Goonies."

I'm a movie fan of that fixed era. Not so much of a strange pastime for most, but for a guy who doesn't even own a TV, it can be a bit of a conundrum. The reason is simple, I love history. Films to me, whether fiction or nonfiction, are portals to the past and so give a unique snapshot of the people, places and issues of a given period.

The Silver Screen of the 1950's and 60's for example was hugely influenced by society's post war obsession with masculinity. So understandably, values then were cast by rough and tumble men like John Wayne and Clint Eastwood in Western Classics. By contrast, in the 70's it was sex appeal from actors like Burt Reynolds, Tom Selleck and Jane Fonda that went tit for tat with the espionage of the Cold War.

Then came the enigmatic 1980's, a period that was more consumed with pop culture than burly heroines. A decade of peace abroad meant youth developed carefree, and without the burden of military service on their backs unlike generations of the past. The result was extreme indulgence that also marked the end of traditional family values as we knew them. They were replaced instead by materialism, fast food and the emergence of the tech culture portrayed in some of my favourite films; "Ghostbusters", "Back to The Future" and of course "The Goonies."

The adventurous tale was of a treasure hunt by a gang of do-gooding kids, who after eluding authorities, skeptical parents and a number of close encounters with bumbling villains manage to save the day, and the fate of their little town, by recovering the lost bounty of One-Eyed Willie's gold. Oh, bring me back the innocence of the 80's any day in lieu of the terrorist bomb plots we're inundated with today. It's safe to say I enjoyed stumbling around a place that as kid had encouraged my dreaming mind.

From Astoria I jumped on the 101 and was intent on a good Coastal Cabotage all the way to L.A. The 101 Coastal, or Pacific Coast Highway as it's called in California, is a dramatic stretch of secondary that runs north to south for more than a thousand miles along the entire length of the West Coast. Winding and undivided, the route links hundreds of communities with enticing names like "Dunes City" and "Gold Beach," and is most famous for steep grades, sharp bends and ocean vistas that divert with alluring rock towers jutting off in the distance.

Before I left Canada, friends had encouraged me to "just take the Interstate...it's the easiest and quickest way." They couldn't understand why I had insisted so strongly on the more iconic secondary routes like 66 and 101. So my response was always to explain it was Americana I wanted to see. The 101 for example is not a drive one takes if in a hurry. Vancouver to Los Angeles on the Interstate is little more than a monotonous twenty-hour drive that by-passes everything. Hardly comparable to the two weeks I spent working the same swath via smaller roads that allowed me to get lost and found in places like Seaside daily.

I had pulled into town to collect Sea Dollars, but drawn to the salty breeze, decided to change and grind out an inaugural barefoot run down the beach. Keeping to the water's edge we jogged for about an hour, stopping often to take in the sights, sounds and energy we found. The tides were near deafening now

The Oregon Coast And Border Delays

and in general seemed to be growing taller with each southbound day. Kneeling before them I enjoyed the anticipation of waiting for cold currents to rush across my feet and bring relief to burning soles made raw by miles of exfoliating sand behind me. The jaunt was gratifying, and as I drove away it lingered because I realized I had invited a new addiction. From that point forward, barefoot running became the norm on every coast I trespassed on.

By the time I neared Brookings I had been exploring Oregon for a couple of days. The town itself is not far from the Border with Northern California, but with night falling fast, I decided to set up camp before it was too late to find a good spot. The ideal site presented at a wide section of shoulder on the side of the road. Pulling in it appeared to have enough room for the truck, and as a bonus, I could hear the tides crashing below. It felt like any one of a thousand rest areas I had passed so far.

It was 9:45 by the time I was done unraveling my gear, and I decided to dig into a six pack of Astoria Brewing Company's finest I had chilling in the cooler. For the first time since I arrived stateside, the ominous grey clouds that had shadowed since the Rainforest were giving way to clear skies and a perfectly starry night. Pulling a beer out, I hopped up on the tailgate, cracked it, and raised it to the heavens to say, "Cheers, Big Guy!" then pounded it back to the chorus of ocean's waves battering the rocks.

The beers were going down like water and with my limbs and body tingling warm now, I succumbed to the whole Pacific

Northwest experience. "It's paradise here," I voiced, but at the same time wondered why motorists kept honking as they passed me. The intermittent noise was a bit annoying after a while, and I recall looking to Athena for council.

"Why are they doing that?" Shrugging it off, I burped as I polished a third wobbly pop, then crawled into the bag, fully unaware that I was causing a problem.

The honking persisted through the night, and although Athena was still able to snore till sunrise, I had been lying awake for some time. Slightly envious of her slumber, I nudged her and she groaned.

"Oh, I'm sorry sweetheart...did you rest okay?" Her heavy eyes and bobbling head were proof that she did. Both of us up now, I rolled to open the hatch, and was pleased to see that even in the pitch black, I had fashioned the perfect spot. "Still got it!" I revered before getting out of bed, but only after I snapped a picture of my feet framed by the open tailgate against cliffs and the Big Blue Beyond.

Taking a break from packing, I stood back and looked at the road to try and figure out what people were honking at. I noticed nothing out of the ordinary, except a steep slope and a blind corner. *"Pretty standard for the area,"* I thought, rationing the act was likely just decorum used by locals to warn each other of cyclists. Dismissive, I grabbed my toothbrush and began to wash up.

Brushing away, I continuing to stand there for a few minutes before I noticed a particular gesture--the motorists appeared to be fingering me. "What is that all about?" I whispered through a toothy grin. At first I waved back, but the measured response was short lived and, probably still frustrated from the sleepless night, I found it much more satisfying to return the favour.

Walking back, I passed the truck on route to the cliff for one last gasp of Oregon breath and shook my head laughing as I did. There was a beautiful wooden sign that looked like art and read "BROOKINGS" behind the truck. Hidden from the road I began to relieve myself of the morning burden of a full bladder and wondered how I had missed the monstrosity the night before. "Probably distracted as I pulled in," I confessed while shaking myself off and admiring the unique hand whittled craftsmanship.

A few seconds passed and a funny inclination struck. At first it was a fleeting thought, but then it stuck. The sensation of a belt tightening around my chest quickly led to butterflies in my gut and images of an imminent accident. I was starting to panic, and as I did my jaw dropped to allow my toothbrush to slip, fall and land bristle down in the puddle of urine I had just left. Realizing what was happening, I squawked, "I think I'm blocking the road signs..."

It all made sense now, with clarity I could see why locals were honking in frustration all night, and that if I didn't get out of there ASAP, I might be nothing more than a headline. With the fear of God pushing me, for the first time in years I remember thinking, *"Move expeditiously, young lad."*

It was term I knew well from my army days when everything was an endless game of hurry up and wait. While as a recruit attending my six month Infantry Battle School Course, I remember the words had certain fear-evoking weight. The Army doesn't suggest anything to its members, instead it dictates to ensure each soldier does what he or she is told. It's a lesson vital to anyone in the profession of arms, and one I had the privilege of learning from a very veteran, albeit intimidating, "NCO" at the time. His name was Sergeant Holohan.

NCO's are the Non-Commissioned Officers of the Service who are charged with getting the job done. If you're a fan of war movies at all, then you'll likely conjure images of guys like Gunnery Sergeant Hartman, from Stanley Kubrick's "Full Metal Jacket." Holohan was that man.

A typically tough as nails Ex-Airborne Paratrooper, he had a wealth of knowledge and a chip on his shoulder. A "real deal" Sergeant though, and by that I mean he earned his Hooks in the operational environment of deployments, not in a training one, his experiences made young soldiers like me want to emulate him. In Army terms, he was all parts.

From Newfoundland I think given the Irish accent, he had red hair, a trim red moustache, and like most Infantry NCO's back then, smoked like a chimney, maintained superhuman endurance, and sported layers of muscle. Not a simple man either, he was also apparently blessed with the gift of gab, because I recall he had one of the most diverse vocabularies of anyone I've ever met. With a calm voice, and in the Queen's English, he would say motivating things like, "Move expeditiously lads, if you wanna eat tonight."

Then seeing confused looks on our faces, he would clarify with language that needed no explanation, and say with a furled up lip, "Oh, my apologies, I can see some of you IDIOTS don't know what 'EXPEDITIOUSLY' means. It means move with a SENSE OF PURPOSE! NOW MOVE! MOVE! MOVE!" The Army's full of real deal guys like him, and trust me when I say he left an impression, and that he set the bar high as far what I still consider a capable NCO.

It had been nearly two decades since I moved as expeditiously as a scared wide eyed kid with an instructor's pace stick about to enter his body. But the skill didn't seem to diminish, nor did the sense of dread that came with it. Within minutes I had Taco piled high with our stuff, and had the pedal to the metal to spin rocks back into traffic. I knew I got away with one, and as I made my way towards the California sunshine, and a reunion with my long lost sister, I laughed hard knowing I would have to reassess the criteria I use to pick campsites, then conceded, "Perhaps I'll do a hotel tonight."

Leg 5
Dogtown And California Bound

"Family is not an important thing. It's everything."
- Michael J. Fox

Although the sky was Mediterranean Blue now, and the ocean breeze called with warm songs of praise to lure my gaze, distracted by the task at hand, I couldn't quite enjoy the scenery...just yet. I was seated safe and sound yes, but right smack dab in the middle of a cliff with my legs dangling some twenty yards above thin air. And because it had been a treacherous tightrope walk down a narrow rattlesnake infested slope to get here, my nerves were still a bit frayed by the effort, hence the reason I was having a hard time relaxing.

To combat the sense of vulnerability, and to bolster my courage, I took a deep breath, then leaned forward and horked an impressive loogie to gage the distance to the churning abyss below. The act was meant to put my mind at ease, but instead it chilled my essence by reminding me that a fall now might forever brand this place unfavourably for my legacy as "Aaron's Bones Cove" or "Foolish Canadian Man's Bluff." Chuckling nervously at the assertion, I questioned, "What was I thinking again?" then whispered the obvious: "I hate heights…."

We had been exploring California for a few days, and in that time I had transited the Northern Shores back and forth, dined with farting elephant seals on a beach where I later slept, and visited Napa Valley for a wine tasting event. Now after daring myself at least a thousand times to do so, I had finally ventured out over the concrete barrier of the Pacific Coast Highway that separates land, sea and sky, to bravely walk where no man should go…or to the untrained eye, shuffle my undignified butt down a slope.

Preparing for the second part of the dare, I reached deep into my bag and past expired road snacks to pull out my last Oregon beer. The can, which was reserved for the occasion, had been chilling on ice before the stunt, so had condensation wrapped like skin around its girth. Cracking it open, immediately I felt my angst disappear with the hissing air, then while somewhat tranced by the panoramic of the Blue Yonder, I began to swig it back showing no mercy. After swallowing hard every last drop, I looked over my shoulder at Taco to wonder how I was gonna get back up, then shrugged it off and instead proclaimed victory with, "EUREKA AARON! YOU'RE CALIFORNIA DREAMIN'!"

I was using a play on the state's gold rush days, but I think I did feel the same euphoria as those early prospectors did when they discovered this region's natural riches. The day earlier was no exception as I made my way through The Avenue of Giants, a stretch of road that's lined by the towering sequoia trees of the ancient Redwood Forest.

Having planted heaps of trees myself as a landscaper, I was inspired by the grandeur, presence and persistence of living beings that were taller than many skyscrapers. But my love for them is more than just a crush; it comes from knowing they are time capsules with secrets to the past. I believe trees store residual energy from antiquity and that if you want to connect to what they've seen, just go ahead and ask with a touch. It's something you'll see me do over and over on the road ahead.

Drawn into their realm on this day, I parked to enjoy a short jaunt among giants whose roots crawled to life before Jesus walked on water. Despite it being midday, the forest floor was dark, but made for an ideal game of charades as I playfully waved my hand through the narrow beams of light cutting through the canopy. The mystery of the woods as always came in tidbits, that is until I stumbled by accident onto the enormous Grandfather Tree.

At twenty four feet across, his trunk was more than impressive, but it was the rings inside that I wanted to touch base with. I was sure they harboured energy from the nuances that included being witness to events like Spanish tall ships sailing passed, and of earthquakes that have rocked this spot many times. Before I left I couldn't help but steal a hug from the old man to charge my batteries with his wisdom.

On the road again, I noticed the landscape was changing rapidly. For weeks it was lush rainforests that gave way to pockets of redwoods, and now transitioned into vast farming deltas surrounded by arid sun baked hills covered in stunted shrubs. The scale of the agriculture was beyond comprehension and everywhere I looked there was an army of farm hands, bent at the waist and labouring in the hot sun to harvest the ground crops of a hungry nation. It was a charming segue to Alameda.

The city is just across the Bay from San Francisco and is where my sister Natasha lives with her partner Emily and infant son Elliot. Arriving around the supper hour, she greeted me on the sidewalk to flag me down after I got lost in Oakland. There we hugged, laughed and spent a few minutes catching up on years apart. After grabbing my bags from the truck, I stopped again to take everything in and made a quick entry into my journal. It was the 12th of December and day thirty one of the trip, the entry, although simple, seemed to capture the emotion. It read, "Soooo thrilled to see Stash, Elliot and Em, there is warmth with them."

Growing up like most siblings, Stash and I had a rocky start. She and my brother Nick always seemed to get along better, and so in my mind anyways, left me out. Real or perceived, I kept to myself and didn't see value in family for some time. As I grew though, I came to appreciate the rare bond that exists between siblings. Friends may come and go, jobs too are often fleeting, but the kind of comfort one feels when with family is akin to the joy of dancing in the dark. Today I see intense worth in family. Its very guild I liken to a castle, reinforced by mortar, rebar and stone. Collectively the structure is sound and provides refuge, but it's made stronger with each testing experience. God knows we've had our share of those.

Waking up the next morning on the couch, I was still riding that high. With a grin I noticed I was surrounded by the remains of shredded wrapping paper left where Elliot had ravaged the gifts from Alberta.

I noticed I had missed messages from a dear old friend named "Ironman Rich." Rich is an actual Ironman athlete who lives in San Francisco. After seeing on Facebook I was passing through, he hollered to see if I had plans. His timing was bang on, and I didn't realize he was in the area, so I was both surprised and pleased to see him reaching out. Always wanting to catch up with an old friend, and knowing full well that locals make the best ambassadors, I messaged him back and we agreed to meet up at the Golden Gate Bridge for noon.

Rich had left Calgary at an early age, even before school was done. Even still he was a proud Canuck through and through, but because he spent so much time in the States, his mannerisms were Americanized too. It took me only seconds to pick his mug out of the crowd, despite not seeing him for twenty years. Walking towards him as he sat waiting on his tailgate, it was hard not to envision the young kid that he once was. On seeing me, he stood to greet me with a smile, revealing as he did that he had filled out well and still carried the same confidence I remembered. We shook hands, and to break the ice I said,

"Time's been good to you buddy." His response was a smart-ass one to mock how tall I got.

"Not as good as heights have been to you man," he said in return.

After leaving the bridge, we jumped in his truck and angled north for a run. He took me to the Redwood Forest on the edge of town, there we did an intrepid six miler. The trail was a confining dirt path offering challenging, hilly terrain that looped like tangled yarn through the trees and awoke my senses with the smell of cedar mulch. When finished, we drove downtown and parked to grab a couple of rounds of Irish coffee at the Buena Vista Pub before tackling Fisherman's Wharf. Irish coffee is a mix of hot coffee, whiskey, sugar and cream used to warm the blood.

A bit tipsy after a couple, we walked the short distance to Pier 39 for a history lesson. There we had a feed of bread bowl chowder and a plate of calamari to soak up the booze before strolling further into the busy market. The zeal there was contagious, and I enjoyed watching the colourful transactions between fishermen, vendors and consumers as sea creatures of every shape and size were tossed, scaled then wrapped in newspaper for waiting mouths. "Never seen a fish market before," I uttered to Rich to remind him of our dry Alberta roots.

From there, we jumped on the cable car for a few hours to ride the open air trolley up, down and around San Francisco's narrow European styled streets. Lined with colourfully painted row houses, unique architecture and perfectly groomed green spaces, the city's transit system made for an ideal people-watching platform. After touring, Rich dropped me off at the truck to conclude the day. Before he left I expressed my gratitude for his kindness and made him promise if he's ever in Calgary again, that he let me return the favour. He agreed, gave a fist pound, and drove away. I walked up the top of the hill overlooking the Bridge and sat lost in the city lights for an hour before also calling it a night. Rich is a good man for taking the time to make an old friend's day.

The next morning was cool with dense fog, and as I stood antsy with the fam at Pier 33, waiting for the ferry to Alcatraz to come in, I remember feeling a sense of irony. Turning to Stash I said, "Never in a million years did I think I would need a sweater here!" It was a naive assumption, but I recall that some years earlier, after reading a quote by Mark Twain that said, "The coldest winter I ever spent was a summer in San Francisco." I had mocked the poor misfortunes of Californians who don't get enough sun. Now as I stood there in the flesh, and chilled to the bone, I couldn't help but mutter the word "Murph." "Murph" being a reference to "Murphy's Law" which states loosely that "if it can, it will." For years it's something people have heard me croak when irony pokes about.

Alcatraz Federal Penitentiary, or "The Rock" as it's called, is a former prison that operates today as a museum open to the public. An isolated island located in the middle of San Francisco Bay, the prison was for decades touted as the feared, inescapable, and inhumane home of America's Most Wanted. After boarding the ferry, I quickly grabbed a hot coffee to warm up, then walked to the front of the vessel for a good view. Once there, I jockeyed among the other tourists patiently hoping The Rock would appear through the fog. Thankfully the weather in San Francisco changes by the minute, and as the sun climbed, its heat began to cook through the haze.

The retreating screen exposed endless amounts of shark infested chop, and as we bobbed forward in it, an eerie hush was consummate among us. All eyes were now focused on the foreboding sight of the island's ghostly cluster of abandoned buildings, jagged rocks, and watch tower materializing in the distance.

Once to the dock I could almost sense the emotion of dread locked within. The facility is intimidating, and I suspect hardened criminals elsewhere, but "newbies" arriving to Alcatraz, felt that same angst as they walked off the dock and looked to the heavens for salvation, only to be greeted by the tormenting sounds of a thousand frenzied seagulls circling above.

I have been fascinated by Alcatraz my whole life, not only because of its rich history, but also because of its reputation as a fortress of fear. Its extremity was America's answer to a crime wave that paralyzed the nation in the 1930's. Over the years the jail has played host to a myriad of the country's worst offenders. Inmates like armed robber "Machine Gun Kelly," Irish Gangster James "Whitey" Bulger and of course bootlegger Al "Scarface" Capone all walked these hallowed felon halls. So dangerous was

The Rock that even respected criminals like them found no safe quarter in a prison saturated with violent men.

People like renowned author, intellect and psychopath Robert Stroud, also known as "The Birdman of Alcatraz" were rampant here. Stroud, who loved birds deeply, had a criminal calling card that included, among other convictions, a series of grotesque murders perpetrated against both inmates and prison guards. So terrifying to others was the Birdman, that he was condemned to spend forty two of his fifty four prison years in solitary. As I walked through the unit, it seemed custom built to contain him, and I wondered if this man wasn't the real life inspiration for Anthony Hopkins portrayal of fictional serial killer, Dr. Hannibal Lecter. A nip ran up my spine as I visualized Stroud staring back at me in the dark.

After breaking away from the tour, I strayed alone for some time and often felt goose bumps develop on my skin via waves of cold air that coursed past to convince me I wasn't alone. Like the prisoners who dwelled here, I felt watched at all times, jumpy even to the point where the sound of a pin drop could have sent me running. The setting was as haunted as they come.

When I bumped back into Stash in the exercise yard, we giggled at my absence. Then seeing the entangled look on my face as if I had seen a ghost, she teased with an antic gaze of her own. From the yard we moved to a large common area with barred windows and a sweeping view of downtown San Francisco and the Golden Gate Bridge. The clear sky was prevalent now and standing there I was compelled to lean forward to put my hands on the window ledge where I'm sure Inmate #85 once did.

"How often did 'Old Snorky' stand here missing the bright city lights he owned in Chicago," I queried, referring to the Kingpin, Al Capone. Although I didn't know it then, I would cross paths with him many times in the months ahead.

After the tour Emily dropped off Stash and I at the shipyard to tour the USS Hornet, a retired U.S. Aircraft Carrier with lots of cool Apollo Space memorabilia. From there we walked back along the shore then made our way down a jetty that was a mile or so long. Once to the end we used the timer app on my phone to capture some great pics together. It wasn't until days later though, while going through them, that I saw how beautifully symbolic the frames turned out. With the sky turning a brilliant shade of purple and the sun setting behind us on the glistening bay, our near black silhouettes revealed smiling faces and an open armed posture that read, "Until we meet again!" It was the perfect conclusion to a long overdue reunion.

By morning I was heading south again, and after several days away, I felt exultant to again forego thoughtlessness in exchange for the addictive serpentine bends of the Coastal Drive. I stopped often on the stretch to have picnics and watch surfers chase waves at beaches with no names. On them I found some of the most interesting people I've ever met, and I was never far from a good conversation or a cold beer with the "young at heart" hippies who call the beaches I was squatting on, their home.

* * *

The temperature was rising, and the slow flow of traffic suggested that we were nearing the sprawling metropolis of Los Angeles. By the time I entered the city we were moving at a snail's pace to the stop and go it's famous for. I smiled though, thinking that on this day I wouldn't have it any other way but to experience the proper gridlock of LA. Continuing, I scrambled to find my camera to capture the moment I passed under the highway gantry with the familiar green overhead sign that reads, "LOS ANGELES," and as I did also paid homage to the thousands of dreamers it had welcomed prior.

The hustle and bustle of the big city was all around and although the agonizing sounds of horns played like insults against the backdrop of yellow smog, the noise did little to obscure the polish of each palm-lined street. I think I may have been the only one with that sentiment though, because as I basked in the opportunity to sightsee afforded by the slog, I noticed my fellow commuters weren't so enthusiastic. Most in fact were slumped over in their seats with their heads propped on their hands as exasperated expressions contorted their faces in scathing ways.

Initially I found it strange and baffling that they seemed so unhappy surrounded by all this, but then I pondered, *"Is the joy of living the California Dream just that? A dream?"* It occurred that perhaps these people share the same struggles we all do, despite the misleading, misconception of flawlessness. Their discontent was proof that in many ways, we're all cut from the same cloth, and subjected to the same grind that steals time, dreams and youthful enthusiasm away.

I found a hotel room in Santa Monica and bummed around town for a couple of days doing the tourist traps of Koreatown, the Chinese Theatre and Beverly Hills. But my pleasure has always come from trying to get lost, and on my third day in LA, I was after just that.

I decided to bang out a long run to see what I was missing out on. It was a hot day with temps in the mid 90's, but with juice to burn, and running at sea level, I handled about eight miles from the Pier and back with no problems. And just by luck I passed a number of iconic spots along the way. Stumbling first onto Muscle Beach, I held my own as I walked by Gym Rats who looked like stacked balloons lifting barbells heavier than me while texting at the same time. Continuing back towards Santa Monica Pier, I came across an area lined with graffiti tagged palm trees, quaint shops, buskers and vibrant verve. Curious of the location, I approached the first local I saw, an older black gentleman draped head to toe in purple. He had gold chains and was the image of eccentricity.

Watching him for a moment, I reveled at his passion as he danced to the beat of early 80's hip hop blasting from a boom box mounted on his shoulder. Seeing my affliction, he waved me over, then with an American accent that was Urban inspired and complete with slang, asked my name. I told him "Aaron" then fired back one of my own, "where are we?" His response was a boisterous one, and through laughter he made me feel inadequate, "You're at Venice Beach maaan!"

"Of course I am..." I thought with a grin before moving on again.

Halting soon after, but this time at the concrete skate park near the water's edge, I became infatuated by the acrobatics of the skaters there. Fearless nomads of the space between earth and sky, their tricks, kicks and flips were dazzling to the eye, and seemed contrary to the laws of physics. Watching their mannerism I could see that these guys were not like skaters back home, they were artists on a totally different level.

Being a wannabe skater myself, but years removed from a sport I never really knew, I was intimidated but compelled to approach, though just to observe for a spell. I found a spot on a small knoll in the shade of a palm to watch the action unfold near a gaggle of people who were sipping on bagged beers as they aired their dirty laundry. A few minutes after I sat down, a younger guy took the open spot beside me, tapped me on the shoulder, and signaled with his hand for a light in exchange for a cigarette.

Instinctively, I grabbed the smoke, flung it in my mouth then dug deep into the pocket I didn't have in search for the Zippo I haven't carried for a decade. Smiling that it couldn't possibly be there, I shook my head at the impulse and gave the smoke back while saying, "Sorry bro... I quit a long time ago." In my mind, though, I scolded the near relapse with, "What was THAT?"

For years I had either smoked a pack or chewed a tin of tobacco a day, and always enjoyed a cigar with each beer. Nicotine was my vice then, both loved and hated, it was an expensive, unflattering habit that took dozens of attempts to kick. The atmosphere of the skate park had stirred a youthful desire to indulge, but I knew even the occasional puff now would tempt fate for a man with my history. I pushed the craving aside, knowing full well that like everything else, it would pass.

"Ya right….me neither," the skater rebutted with a smirk on his face as his friend leaned in with a light to initiate a long hard pull of toxins. Then exhaling through closed teeth he said as if trapped by it, "Soon anyways...I hope…" The comment brought a shared commonality and we engaged in small talk. His name was Cartwright, and I remember thinking as we spoke that he looked, talked and acted every bit the part of the So-Cal skater I had envied growing up.

Tall, skinny and with hair shaved close to the skull, he wore long black denim skate shorts that hung low on his shirtless, tattooed body. On his feet were red and white striped knee high's sprouting from green Adidas Kicks that rested lounge-like as he spoke on a well-used skateboard doubling as a foot rest.

We both opened up and I soon found out that although we have very different lives, we also could relate, namely because we were transients from some other place. The City of Angels, I was learning, is really a city of strangers. I told him about my trip, how I enjoy longboarding, though I'm no good at it, and conceded that there's no way I could ever keep up with these guys. Smiling at the admission, he forced over his most prized possession and said, "Don't sell yourself short...give Dogtown a try." He then directed my attention with a nod towards the fray.

I scanned the area. It had looked familiar, but I couldn't put my finger on it until just now. This was the backdrop and shoot location for the popular true story of where skate culture originated called "Lords of Dogtown." Despite knowing I might embarrass myself, I took him up on the offer, grabbed the skate, then spoke in proverbs, "When in Rome..."

Placing the board down, I took control of it with one foot then stared out into the crowded space for several minutes. The place was as packed as a disco, and I figured I could, at the very least, blend into the crowd for one or two passes without doing anything too stupid. Not knowing the etiquette I was apprehensive, but threw caution to the wind and pushed my way in.

Moving slowly and like a fish out of water, I began to loosen up until I had a false sense of security. With tidbits of momentum, I passed over smaller contours of the concrete surface, but dared not to cut and weave aggressively like the others. I felt clumsy and could almost sense spectators giggling at the sight of an awkward old man in running shorts that looked like hot pants cutting through their sacred shrine. Pros added to my angst by zipping by at breakneck speed fast enough to wobble me. But as I finished my first easy crossing, I didn't care because I knew I would never see any of them again.

Paused at the edge of the park with the board kicked up under one foot like a veteran, I felt strong and upped the ante by picking a bigger prize; the smaller swimming pool like feature on the far side. It was only about a three foot curve, but judging by the demographics over there, I'd wager it was for children, not men. Still something about that curve seemed alluring and called my name. *"Surely I could tackle that thing?"* I thought in jest with the confidence of a peacock.

I took a moment to think about it, then as before pushed off, but this time with pep. Faking it as best I could, I assumed the position I had watched others take before they went airborne by bending my knees, rolling my shoulders forward and hunching into a lower profile. On entering the curve I could feel the gentle slope of a new grade develop beneath my wheels and as I transitioned into it, I realized I was fully committed now. No big thing though, the ramp was child proof, but I did began to wonder if gravity was wanting to have a word.

This was a theory I shouldn't have summoned, because barely up the curve I was certain now that it was pulling me back to Earth. It was at about this point that I inferred something I already knew, "I'm not built for board sports because I panic with the unknown." The slight detection of G-forces creeping in felt alarming, and it was enough to rattle my conviction. I began to teeter awkwardly in search of balance I never lost, and soon I fell backwards to land flat on my butt.

I hit hard enough to knock the wind from my chest, and just in case not everyone saw that, the chariot I had rode in on was now sailing through the air. Embarrassed at the gaff, I laid there motionless for what seemed like hours trying to catch my breath and wondered if things could get any worse. Just then, and only by the rule of "Murph," my board rolled back down the incline to nail me hard in the ribs. *"Bailed,"* I thought with another big gasp.

With laughter and the sound of clapping bearing down on me, I raised my head to find its source. Cartwright was approaching with the same encouraging dimple he used to send me out. Arms open and looking to the sky he vocalized "NICE!" then pulled me to my feet and patted me on the back.

"Nothing gained…nothing lost…good show!" Breathless still, I tried my best to play it cool and accepted his praise.

It was early afternoon when I left Venice to head back to the hotel for a quick shower and to change out of my running gear. I asked Cartwright if I could buy him a beer later to say thank you. He said, "Yup…I'm around", but when I returned he was nowhere to be found. So I just walked about for a while. For dinner I sat on a patio overlooking the Strip where I had a burger and nursed beers while watching surfers chase their last waves of the day against the backdrop of Santa Monica's Ferris wheel.

Within moments the sky above turned black until only a ribbon of orange existed to highlight the vast horizon leading west. "What an incredible place to grow up," I acknowledged. LA turned out to be interesting; the weather was mild, the locals friendly and the activities plenty. The next morning as I drove east away from the fertile fields of heavily populated California and into the open moonscapes of desolate Nevada, I felt excited that just down the road lay the infinite space, beauty and mysteries of America's Southwest Desert.

Leg 6

The Sandy Coulee Empire Of The Southwest

"If California were pretty, then the desert is the image of masculinity." - A.L.

Leaning against the front grill of Taco with my arms crossed, I watched amused as heat shimmered off the surface of the asphalt, then danced across the thirsty earth with a seducing, almost mesmerizing effect. The landscape was different now; red, desolate, barren, and baked by the relentless sun, it felt unforgiving, but provocative too.

Pulling the beak of my cap down low to deflect glare, I examined the dirt beneath my fingernails from nights spent camping beside giant cacti, and smiled wondering if I could even count on that same hand the number of vehicles I'd seen pass in the last few hours. Then looking back over my shoulder at the spinning blue cherries of the Highway Patrol car parked behind me, I laughed to mock my own misfortune and said, "Actually...I think that's the only one."

Moments earlier, he had been set up behind a large boulder on one of the countless blind corners I had careened around that morning when he tracked us doing sixty nine in a fifty five. After pulling to the shoulder, I watched through my mirror the officer's hawkish demeanor as he exited his white cruiser, gave a big two fisted heave of his pistol belt, then with a sigh, pulled a pen and notepad from his breast pocket to approach.

I waited for the tap, and when it came I rolled down my window to greet him. The instinct was to be cordial and so I tried to offer an apology, but he beat me to the punch with the bluntness of a twitching moustache and a half polite, half sarcastic, "Do you know how fast you were going?"

I really didn't, but for the sake of argument and to expedite the whole process I weighed my options then responded in jest, "Not nearly as fast as I had been going before I saw you."

I could see the admission took him by surprise, because with it his guard dropped to reveal rolling eyes and a slight toothy grin beneath that monstrous cookie duster of his. He began to lead the conversation with talk of safety as he wrote notes and subtly prodded for details of just exactly why I was literally flying through his parched neck of the woods. In turn I revealed my grand trip and the events in my life that lead me to him that day. Then once I finished explaining, I asked his pardon again for the speeding, but confessed also that I had been doing so pretty much since I left Los Angeles a week ago on the hypnotic, nearly abandoned, always iconic, American Route 66. Based on his approving nods, I gathered he had heard the story a time or two before.

After grabbing my licence and registration he clicked his pen back into its sheath with his thumb then put it back in his breast pocket, and told me he'd be right back. Twenty minutes later he returned with my papers in one hand and a folded ticket in the other. Looking me sternly in the eye, he handed them back and said, "Enjoy your trip, Aaron," then turned to jog back to his patrol car where he got in, slammed the door and sped off.

"Weird," I thought, figuring that he must have had another call. Opening the ticket to take a look at the damage inside, I giggled at his lenience. In large block letters and written in still wet ink, it had a voiding line drawn through it, but read in lieu, "WELCOME TO ARIZONA, SLOW DOWN PLEASE!"

Welcome indeed. Deserts for me have always felt like home, and I had looked forward to the Southwest leg since even before I left Calgary. Having grown up not far from the skin cracking dryness of Alberta's own Badlands, I've always had an affinity for their arid beauty. And it seemed that the affection was now paying dividends with good luck, because the run in with the cop was the second close call I had.

A few days earlier I had been camping up North in one of the many Canyons I had pulled over to photograph. While star gazing that night, I got up to top off a drink and my Canadian themed chair, whose name is Rosy by the way, also got up and floated away. By that I mean a gust of wind had taken her, and although I could hear the frame clamor somewhere near, with a moonless sky, and darkness as black as a basement with no lights, I knew there was no way I'd find her. *"Oh well,"* I thought, as I took it as a sign to go to bed, and resolved to resume the search in the morning.

At first light, I got up and walked over to take a gander over the ledge she tumbled, and scoffed hard at what I saw. Standing completely erect and even facing me to project pride with its Red Maple design, was my ambassador, Rosy the chair. "Murph," I smirked, wondering what the odds of her landing on all fours were.

Hopping from one boulder to the next, I began to make my way into the coulee to rescue her, but was stopped about halfway down by the strength of intuition. My spidey senses were tingling, those are the "sixth sense" that usually warn the eyes have seen something the mind hasn't yet. In the moment my intuition was telling me there was something near the chair. Leary, I strained through squinted vision until a clear picture began to emerge.

"You sly bastard," I thought as I considered whether or not I really needed the chair moving forward. It was an appropriate question, given that just abreast of it, and guarding it like a prize, was the body of a cocked, loaded, and ready to strike, little rattlesnake.

The serpent lay nearly motionless as I arrived to the bottom with a clear case of the collywobbles. I've always enjoyed catching garter snakes back home, finding the texture of their sleek, scaled backs deceptively inviting to touch. But with a sinister bite, I knew this guy wouldn't appreciate that, so I kept a bit of distance between us. Within a few feet now, I remained cautious and smiled while kneeling and offering words to relate, "You look almost like the sunning rattlers I've met in Alberta."

Unimpressed by my name dropping abilities, he stared back with a cold reply and tasted my scent on the air with his flickering black tongue. Not wanting to test his patience, I admired him only for a minute before reaching slowly for the chair and asking rhetorically, "Surely you don't mind parting with this one item." I snatched it away as fast as I could.

Once I had it in my possession, I ridiculed my apprehension because this snake, I was sure, could neither fly, nor run, nor would he want to. In fact he probably doesn't even care for chairs so why would he take exception. After apologizing for the interruption, I made my way up the slope again, then turned back at the top for one last look. He had slithered into the rocks.

The chairs perfect presentation lingered in my mind as I drove away and conjured a memory of a time, while serving with the Army in Afghanistan, Willy and I affectionately did nearly the same thing to our old buddy Tom Hamilton, or "Hami" as he was called by friends. The three of us, like the Stooges, were tied at the hip from innumerable experiences shared over the years, like the tour we did together there, and so practical jokes were inherent.

On this occasion we had hauled all of his kit; sleeping bag, cot, barrack boxes and everything else he owned, into the middle of our ball hockey field, opposite the sandbag castle we called home.

Once there, we set it all back up to resemble the exact same layout he had inside, even going so far as to use a ruler to measure his bug net, boot alignment and the placement of the pictures he had levied about.

It was a painstaking endeavor, but the look on his face after was worth every bit of the effort. Hami wasn't laughing though, and in fact he seemed downright irate about the whole ordeal. Willy and I still joke that never once after did he even give us a pat on the back for the hard work it took. Fortunately though, his sense of humour was sound, and although we knew he would strike back somehow, for now he relaxed and showed his class by simply insisting, in not so many kind words, that we best help haul his kit back in.

It's one of the many imprinted memories from a place that really wasn't all that nurturing to begin with. The Canadian Army in Afghanistan suffered thousands of casualties in one form or another, so those rare occasions of forced laughter were the escape we needed from reality. Plus they were the lesser of two evils when you consider our other pastime was to cheer on American Bombers flying low overhead on route to drop their payloads in the mountain passes. A nightly show for us to watch from observation posts on the sides of those mountains as they lit up the sky with enough flash to make thunder and lightning blush.

The Islamic Republic was once the cradle of society, but for generations it has been contested by many sides. Her rugged mountains, deserts and vine-lands harbour angelic, often nomadic people who are humble, kind and touching. But if pushed to the edge, are as ferocious as any warrior culture that has ever walked. When you look into the eyes of a Jihadist Warrior for the first time, you'll see what I mean. They are fearless and piercing, and no words could explain to the know-it-alls back home who give their two cents on how to win a war they'll never fight in, that you can't beat people who aren't afraid to die, it's just that simple.

It is a country of contrasts, so hostile, yet so enchanting that when you're there, you count days down to come home. Then with addictive paradox when you're home, you count days until you can once again return to bathe under the bluest sky God ever made. Of course for a twenty something serving on a NATO tour there, it doesn't feel so inviting; in fact it felt outright deadly. The life of an infanteer is always one of hardships where blistered feet, sore backs, cold food and sleepless nights dominate the routine. But in that place, even those inconveniences can't distract from the certainty that any sunset might be your last.

The work environment is nerve racking, and with heightened senses constantly being tested, it's not uncommon for a single tour to visibly age a person's mind, body and soul ten years as they battle, then and now, the sensation of sand storms, dehydration, and close calls, never mind habits that will forever make you tap your boots for scorpions in the morning. It's fair to say that the American Southwest Desert leg was summoning all manner of memories I had forgotten. And although I was looking forward to catching up with Willy and Hami once I got to Halifax, which was still a half a continent away, I also felt a bit of trepidation about the meeting because it had been so long, and because I knew we had a lot to talk about.

* * *

The four corners region of America's Southwest comprises Utah, Colorado, Arizona and New Mexico. Intent on seeing as many of those diverse landscapes as I could, I enjoyed zig-zagging across their boundaries carefree. Already I had spent a day hiking past cascading waterfalls in Utah's Zion National Park, walked the rim of the mile wide Meteor Crater and visited Hoover Dam after a night in Las Vegas. Now I was headed to "The Antelope" for a spiritual stole.

Antelope Canyon is a protected site that's fiercely guarded by local natives. To enter, one needs not only permission from the tribe, but also the employ of one of their qualified guides. The strict process ensures that people don't squat in the caves, but it also guarantees their safety. Guides will remind you that the canyon was furrowed by rainwater, and that just because the sky above is blue, doesn't mean the risk of flooding isn't there.

In the region, storms develop dozens of miles away and well upstream of the canyon, where they dump massive amounts of rain. Then with little or no warning, the gathered tide is thrust through the systems narrow catacombs, and like the swirling drain of a giant bathtub, soon the entire basin is purged of everything. By the time you see or hear the desert tsunami coming, it's already too late, because there is no escape.

Being washed away is a scenario that seems unlikely for most, but the phenomenon is actually quite common to The Antelope. In 1997, eleven tourists, most of them from Europe, were killed when a flash flood swept them and their ladders away. It was a fact that weighed heavily on my mind as I passed through, and I made sure to pay my respects as I would at any final resting place.

Despite knowing the dangers as I walked around, I took my time and stopped often to collect my thoughts as I attempted to dissect some of the most pronounced art I have ever seen. The walls were carved as smooth as pottery, and had varying shades of orange, dispersed through thin layers of sediment that appeared to be decorated with deep red undulating lines, painted, I imagined, by the care of an artisan's hand.

Stumbling in the soft eroded sand that collects at the base of them, I gazed high to the ceiling above where only sporadic slivers of light were permitted to breach the Canyon's great depth. Those that did created an impressive panoply of colours on the rock surface, where without thought, I began to run my fingers over them, tracing playfully as I did, their shapes until I realized my tips were weathered as raw, smooth and shiny as the sandstone acropolis itself.

"This is not a place for the claustrophobic." I reckoned as I exited the cavern and walked back to the guide's truck. Hunkered down on the back bench of the open boxed vehicle, I struggled to keep dust from my eyes as I watched, fulfilled and elated by the experience, the canyon disappear. It was December 24th, Christmas Eve, and with so much still to see, I was eager to conclude the day by getting back on the road. After being dropped off at Taco, we drove late into the night.

Waking at dawn, we found ourselves the next day back on the Utah-Arizona border where I could see in the distance the impressive towering Buttes of Monument Valley. *"This is the place legends are made,"* I thought, while admiring the rock formations that jut more than a thousand feet in the air and bring life to an otherwise desolate plateau, guarding without condition, the surrounding Navajo Lands.

Parked in front of them now, I grinned as I watched the sun dance, then crawl across the sky, ducking in and out of the Buttes presence like a small child. Everything looked so familiar, and with warmth flooding my body, I wondered if I wasn't in the exact same spot Hollywood cameras used to track "The Duke" galloping heroically across Grandpa's fuzzy old TV screen where we gathered as children on pillows to wave him home.

Wanting to head out on a run to connect to it, I observed the scrub land for a moment while changing and lacing up. As I did, I realized every bit of stunted brush and shadow out there could conceal dangers in the form of basking rattlers like the little one I encountered earlier in the week.

Knowing that Athena's natural curiosity leads her astray, I convulsed recalling a time in Calgary, when on a run, she bolted past me, and to the horror of elderly folks sharing the path, showed off the squawking grouse she had caught and was carrying in her mouth.

"Sorry, sweetie...this time you can't come." I led her to the truck, rolled down the windows and headed for a long jaunt down the Wildcat.

The Wildcat is a charming 3.2 mile stretch of largely unmarked dirt trail that circles the East and West Mittens of the park. Enjoying the route immensely, I ran it three consecutive times, stopping often on the first loop to find my bearings because I was lost, on the second to photograph the scenery, and on the third...well... to confront the two reserve dogs that were chasing me.

I've been chased by dogs before, but none as determined as these ones here. On the first two loops I noticed the animals patrolling the edge of the community, and so simply picked up my pace to get through the area. The strategy seemed to work, and at first anyways kept the odds in my favour by ensuring a good time distance gap between us. By the third pass however I could see them trotting in my direction with their heads held low as if they were onto me. Still not really all that concerned, I headed for low ground and followed a dried up river bed, certain that by putting myself out of view, I would put myself out of their minds as well. In hindsight the assumption was a bit callow though, because within a few minutes I could hear a salvo of barks closing in as they followed.

Sensing they were near, I stopped to survey again. I was miles from the truck and saw nothing but infinite space in every direction. Images of being torn limb from limb by wild dogs, then left for scavenging vultures consumed my mind. The first instinct was to sprint in the opposite direction as fast as I could, but aware I move faster than nothin' on legs, and knowing such a move would likely trigger an attack, I decided to stand strong.

Planting my feet firmly, I turned my body sideways, like the Karate Kid, then forgetting my predicament, released a burst of caustic laughter wondering where the urge came from. Anyone who knows me knows I know nothing about martial arts, so I'm certain the bluff wouldn't have worked out. The position did bolster my confidence though, and with a clear head I noticed two baseball sized rocks that seemed perfect for my defence. Kneeling forward, I picked them up, and stood just as the mighty beasts crested the small hill that banked the river bed.

Their robust hale, a combination of barking, growling and snarling sounded primitive and rippled right through my senses as they advanced in a cloud of ash so fast, and so furious, I thought the Tasmanian Devil was in pursuit. "Whoa! Whoa! WHOA!" I yelled boldly as they neared with my rocks held high to show I meant business. The effort to intimidate did nothing though, and I knew that if I was pressed to throw now, while they were still out of range, that I would surely miss and risk a good mauling.

"What would Texas Ranger do?" I thought, still stuck on a martial arts fix while referring to Chuck Norris, the man so bad he counted to infinity…twice, then wrestled a tornado to death. Desperate times call for desperate measures and impulsively I did what only he himself would; I began to frantically kick dirt at my

tormentors. With both legs working furiously, and my arms flailing in a globular fashion to counterbalance the assault, I released an awkward jig, and a ton of earth.

Now I'm aware it was hardly the act of a courageous derring-do warrior, in fact it probably came across as completely absurd, but credit to Chuck, it worked. After a few seconds, and out of breath, I relented to gather myself and stood frozen with inquest in a cloud of puffy dust while coughing a bit. In the blinding grime, nervously my eyes shifted to and from as I wondered where the dogs had gone. "HELLO?" I shouted out.

Maintaining the stance, and with rocks still held high above, I could feel sand grinding in my teeth, and noticed a film of red dust covering my sweat exposed skin. Smiling, perhaps prematurely, I felt as though I had dodged a third bullet in the desert and began to relax. Just then I heard a snort through the settling haze and while straining my neck towards its source, realized that only a couple of feet away I was being ogled at by two sets of dumbstruck blinking eyes. Their silhouettes began to take shape.

Covered in the same red dust was the titan black coat of a sitting, drooling, Rottweiler and his bossy companion, a tiny tanned Chihuahua. The Rotty had the muscles and presence of a Bull Moose, but threatened nothing with eyes that were patient and begged the question, "Why would you kick dirt at me?" By contrast the little one seemed obnoxious, loud, and began again to lead the way with aggressive growls to test my resolve. I don't know about anyone else, but if you're like me, when you get threatened or cornered, at first you get scared, then you get brave, then you get really pissed off.

Seeing that the two respected authority, I scraped my foot along the ground to create a small pile for another flail, and while I kept my eye on the little one, I said through clenched teeth, "Back off little man...." Then as he halted to consider my warning, I went on the offensive and kicked the pile into his face before bolting. I could hear him yelp, but with no pitter patter behind me, I stopped a short distance away to make sure I wasn't being followed. As if the fracas were routine, the two were trotting away. Looking to the sky above, I pointed and said, "Thanks, Big Guy."

I had a bad feeling about Athena coming, and I'm sure that was the danger I had perceived earlier. After getting changed I sat on Tacos tailgate and ate a dry pack of oatmeal to refuel. Not wanting to leave, but always on the move, I snapped a panoramic of the valley, and when I got a signal, sent it to Shuby and Murray in Calgary with holiday wishes.

It was so beautiful I couldn't help but reflect on past places I had spent Christmas Day. Where when most people are safe and sound with family and friends, often I've been anywhere but at home. There were years that celebrations of any kind held no value, especially as a young child when there were no presents under the tree. But as an adult who embraces the birthday of Christ, regardless of where I am now, I pause to give thanks and recall the best Christmas I ever had.

When Mom left Dad, she struggled to get back on her feet, and while we had ample support from family, that man certainly didn't make it easy. His goal was always to prescribe chaos with fear and, despite court orders, he harassed and stalked us for years, forcing dozens of moves, just as many new schools, and long stints spent at Calgary's "Discovery House."

The Discovery House is a non-profit organization and women's shelter that takes in battered women and their families for usually up to a year. Quite flexible however, we stayed much longer. The organization offers essential counseling, healing and rehabilitation services needed to empower women and give them back a sense of dignity.

Going to bed on Christmas Eve with no gifts under the tree is deflating and I recall praying that Santa would come, but because of lean years before, I had no expectations he would. That morning I slept later than most and didn't even roll out of bed until my older brother Nick came and woke me up. Stumbling out into the living room annoyed, cold, and wrapped in a blanket, I was overcome.

With lights twinkling, candles lit and the smell of comfort food in the air, the room literally reeked of yuletide cheer. Mom was crying as she sat on the wide armrest of the chair hugging Stash. Tucked between them and the adjacent couch, in the far corner of the room, was an ugly four foot Charlie Brown Tree that she had picked up. Hardly a prized Balsam Fir, it seemed purposely born to be the butt of jokes. But despite our protests, Mom defended it, reinforcing as she always did that there's more to a soul than its flesh. There was no need to defend Charlie this day though, because he was barely visible under a mountain of gifts we won in a raffle organized by the shelter.

Approaching slowly, I drug my feet in disbelief at what I'm sure is probably the same overwhelming euphoria a lottery winner gets. Nick and Stash were already tearing through the pile like prospectors, but instead of picks and shovels being wielded, it was wrapping paper and laughter that filled the air. For the first time in a long time, we felt normal and were able to showcase our loot at school like everybody else. That was the least of what the Discovery House did for us, but the Christmas was the best I've ever had. Monument Valley was a close second.

* * *

The next morning was frosty as I attempted to shake life back into my numb body after a long cool night. Draping my sleeping bag across my lap for warmth, I leaned forward to open the tailgate so I could get my first look at the surrounding Ponderosa Woods that had creaked and moaned all night as wind whistled through their ranks.

Feeling exhausted I knew I was putting in too many hours on the road during this desert leg, some days up to fourteen, and far more than my goal of four to six, which would allow a good balance of rest and play. *"But who could blame me?"* I reasoned in my head. With abandoned roads, constantly changing landscapes and remote camping sites everywhere, I was enjoying the journey and proud of the nearly three thousand miles I had conquered in the desert.

Up and dressed now, I heated some water in the kettle, poured a cup of instant coffee, grabbed a granola bar, then headed down a narrow path through the woods and towards the glowing horizon. As I walked I began to notice brilliant shades of reds, oranges and purples develop as the land of the Rising Phoenix crawled to life. Nearing the end of the trail and giddy with anticipation I paused for a moment then took a deep breath, knelt under the last heavy pine bow and entered the vast expanse. The sight overpowered, and with a sip of coffee and a breath of fresh air to absorb it, I exhaled the words, "Finally, we've reached the Grand Canyon."

I was on the North Rim looking down at the same mighty Colorado River that stripped the land of two billion years of history. In front of me lie the most impressive gorge the Earth has ever known. At nearly three hundred miles long, twenty miles wide and six thousand feet deep, it was a sight to see. Emotionally moved, it was impossible to not realize the significance of that place, and I found myself visualizing the daring expedition of Lewis and Clark braving the rapids below as I walked in the footsteps of the manliest man of all time, President Theodore Roosevelt.

What many saw as inhospitable wasteland, he saw as serene enough to declare it a National Monument. Then he had the foresight to urge Americans to preserve what defines the

Southwest today with, "You cannot improve on it, but what you can do is to keep it for your children, your children's children, and all who come after."

The Sandy Coulee Empire had so far been remarkably lucrative. Its unmatched space seemed to hide everything from awe-inspiring landscapes to close calls with animals and police. And despite the grandeur I had assumed I would find, it was the small things, like the desert coyotes who sang me to sleep each night, that kept me on my toes. I felt I had a connection there and knew I could have easily stayed another week. But with my tank full of solitude, and realizing that true nomads keep moving on, I prepared for one last stop by heading to a place I know gunfighters still stalk.

Leg 7
Outlaws Of New Mexico

"I'm your Huckleberry...." - Doc Holliday

Not far from the Mexican border it was picturesque rolling hills, cacti and scrub ranchero land that now eclipsed, as I stood under the rusted out steel sign on the edge of the old part of town and questioned, "Just what kind of name is 'Tombstone' anyways?" One that spoke to her Wild West past I was sure, and being a proud Western myself, I smiled to embrace it with, "Only in America."

This was my last stop in Arizona, but the first real one in an outlaw odyssey that would see me conclude the Four Corners by pursuing the Gunfighters, Cattle Rustlers and Stagecoach Robbers that made it famous, and that as a young lad had captivated my mind. Tombstone, like most other towns on the frontier during the nineteenth century was a mining boom town, turned ghost town. Today it booms again, but as a custodian of that history after the big screen revival of its legends. We can give credit to the 1993 blockbuster of its namesake for that. It glorifies the exploits of, among others, lawman Wyatt Earp and gunslinger Doc Holliday, both of whom I was there to see.

After parking, I walked into town on an unpaved road following, appropriately, the path of a rolling tumbleweed as it led me past wooden shacks and right to the front door of the original Birdcage Theatre. Inside, I looked to the tiny stage with a grin and couldn't help but wonder what energy it harboured from its more turbulent days. If you need proof it was a rough place to hang out, just look up to the ceiling to count bullet holes that confer like outlaw graffiti. That's what I was doing when I forgot the crowd around me and blurted out, "I'm your Huckleberry," half expecting to see the apparition of 'Doc' materialize before my very eyes, drunk, stumbling, but with a quick draw, still quite formidable.

Heading down the street, I visited Big Nose Kate's establishment before venturing into that infamous horse stable for a reenactment of an incident that defined America then, and ensured the days of her Wild West were numbered, the shootout at the OK Corral. It's an incredible tale, and I enjoyed watching its reenactment by actors dressed in original cowboy garb from the period. With blanks they brought back to life that fateful day in the fall of 1881 when the Earps, Clantons and McLaurys all clashed for control of this tiny burg.

Trying to visualize the state of affairs, I imagined the tension was thick in the air as frightened mothers scurried to grab their kids, shopkeepers locked their doors and even horses, tied to posts, cried out as intuition warned. The spot where the gunfight took place was a battlefield, no question about that, but the smallest one I've ever seen. Once in front of it, I wondered how any of the nine men involved could have possibly got out. In close quarters, the two sides stood only six feet apart, and in a two second frame, released some thirty shots at each other. When the smoke cleared, little was settled, but three lay dead and most others wounded. Aware that such confrontation was the norm back then, I headed to Boot Hill Cemetery to pay my respects before leaving Arizona.

* * *

The next day I entered New Mexico and stopped for a quick run down an abandoned road, where once again, I was chased by a pair of desert dogs. They were tiny this time though; cat sized in fact, and although I could have kicked them off, I didn't have the heart, so let them nip at the heels of my sneakers the whole way back to the truck.

The landscape here was again in constant flux. Gone now were the brilliant reds and oranges of Utah and Arizona's Sandy Coulee Empire. Replaced instead by an inspiring cache of rocky grasslands, painted on either side by the silhouette of black mountains so far in the distance they appeared to be mirages. Feeling like an enthused explorer in a new realm, I stopped often to examine plants I'd never seen, stood in the shadow of boulder outcrops that looked to be stacked by giants, and adored everywhere the magnificent palette of browns and tans that I was sure were borrowed from the lunar surface.

The vastness was impressive and seemed to be made more magnificent by the long linear roads that worked like arrows to direct my gaze as far as the eye could see. In many ways New Mexico's remoteness rivaled the Canadian experience to a tee. The Great White North is one of the least densely populated nations on Earth. And although we have millions of people living in urban centers, by and large the country is abandoned. It's a fact I like to throw around to our American cousins, most of whom could never imagine such boundless expanses.

My oldest best buddy in the world, Bobby Robert Douglas Kendall, or Bob as he'd be the first to remind you, could attest to that, because he's also my fishing partner. Today we are confidants closing in on forty something lifestyles, but we go way back past the bar room brawls of adolescence, and even further into the elementary years of innocence. And with so many shared experiences together, it's not surprising that we have that unique kind of bond that allows us to sit through a night without scarcely saying a word. It's a rare true friendship in life. His family of entrepreneurs owns the Crossroads Farmers Market in downtown Calgary, a fixture in the community. Annually they organize a fishing trip to Northern Saskatchewan, and I've been fortunate to have been invited a couple of times.

With thousands of untouched lakes, it's no mystery why people from all over the world flock to the area. The fishing is first rate. Spoiled by it, we joke that if a man is not pulling his weight with a prized pike or walleye on just about every cast then check the line, because it's likely twisted or broken in some way from a strike.

But it's not the fishing holes I brag about, it's the journey to get there that is truly Canadiana. Leaving Calgary in June we drive northeast for about sixteen hours to the end of the last dirt track on the map. There we arrive to a cluster of log cabins on a lake with no name where we rest for the night. At dawn we put the boats in the water and wade several miles into the backcountry until we reach the opposite shore. Once there, we dismount and portage the boat some distance by labouring it, and everything else we brought, overland and through dense woods until we reach yet another totally isolated lake. Now as if that's not secluded enough, we then wipe the sweat, file into the boat again, and journey several more miles into the unknown, until we finally arrive at, and set up on, one of many tiny islands in the middle of nowhere.

Surrounded by crystal clear water, the mosquito laden islands are frequented only by bears, wolves and their prey who share virgin pines still immune to chainsaws. When there, we fish by day and grow fat on the bounty by night as the crackle of fire competes for attention against laughter, flowing beers and the hale of grossly embellished tales.

That is the Canadian experience, and while driving through New Mexico, I was reminded that we share a few things in common. And if not for one peculiar detail, I might have easily felt at home. That being, that four times in two days I had watched trucks of the U.S. Border Patrol pass in the opposite direction, burn a U-turn, then tail us for a few tense minutes before lighting us up. I'm certain it was Taco's closed canopy that caught their attention, but the unmerited pullovers felt invasive.

I knew that they were necessary though given the security climate along the border. As always, my instinct was to cooperate by being polite and talkative, but these guys would have none of that. Like the guards who tore us apart a month earlier in Washington, they seemed exasperated by their job, and so understandably lacked the joy of emotion.

Querying for information to put my mind at ease was like pulling teeth, and even with simple questions like, "What's going on, sir?" I received only short, abrupt answers that were designed to intimidate.

"We're looking for illegals, drugs, weapons or anything else we deem criminal," they would say. Then, as if to punish me for asking, one would watch me with eagle eyes, while the other would search the truck. Finding nothing, they would wave me off without so much as a, "Have a nice day."

Driving away, I realized they're products of their border town workplace, so it wasn't anything personal. The border towns are notoriously violent places that harbour all kinds of criminals from Mexican Cartels, to arms traffickers, to people smugglers.

Exploited as gateways to the north, the pull overs along them were a stark reminder that in many ways the Western Frontier of today is as wild as it was a century ago when bandits like Jesse James, Butch Cassidy and Curly Bill stocked this realm.

I had been following Route 70 in pursuit of men like them for some days now, when just by chance I came across a site I recognized; the high features of White Sands National Monument. It had been a long drive and tired, I decided to rest before exploring them. Pulling into the park, I ate a quick lunch of apple slices and peanut butter before dozing deep into a neck wrenching sleep in Taco's front seat. I awoke hours later stiff and groggy, but charged enough to tackle the park.

Together, Athena and I hiked into the dunes, then trekked up their slopes and like a snowboarder with a running start, slid right back down to the bottom, where I would shake the sand from my shoes and question, "How did they get here?" The dunes, glistening tall with the sun's glare, were a thousand shades of crystal white and were dotted with shrubs that stitched the smaller mounds together. To me their appearance presented like ocean waves, but ones that flowed four thousand feet above the sea. It was an incredible site, and reminded me of the Oregon Coast, but unlike there, I knew this place had an ominous side.

Walking back to the truck, I turned one last time to snap a picture of them. As I did, I found it impossible not to visualize on the horizon a mushroom cloud billowing a mile high into the sky as waves of radioactive energy, triggered by the blast, clapped their way across the land like ripples through a pond.

"*This is that Trinity place,*" I thought with a chill as I reflected on images etched in my mind of America's first Nuclear Bomb test. It was a dark chapter that played out right on this very site back in 1945 as part of the Manhattan Project. Motivated by the secrecy of that, and perhaps sensing conspiracy, I felt compelled, even obliged, to deviate from the bandit quest I was on, if only briefly, to pursue a good old fashioned New Mexico style cover-up.

Feeling like David Duchovny's character Fox Mulder from "The X-Files," I headed several hours east to Roswell, where in 1947, a supposed flying saucer crashed in the desert and made headlines around the world with theories of counterplot. Driving into the city, I smiled at the tourist trap. With hotels, businesses and even street signs decorated by aliens and sci-fi relics, I could see the communities claim to fame was still their bread and butter. Curious to learn more, I went straight to the UFO Museum and Research center to see the evidence for myself.

Entering I was greeted by friendly staff and a well laid out presentation of chronological events and displays to support their case. Included was information on the initial crash site, pictures of the debris, subsequent investigations and even newspaper clippings that detailed testimony from witnesses. I must say their argument was compelling. I do believe we're not alone in the Universe, and that we will find life somewhere else, but I left Roswell the same way I drove in, a skeptic. I'm not saying I can't be swayed, because I can, but as it stands, generally I believe the Air Forces version that construes the pieces of wreckage found were in fact parts of a classified nuclear balloon. Which is consistent with the history of the Manhattan Project in the area, and explains the hush-hush approach.

Back on track, I was finally in Fort Sumner after several days of following the narrow, shoulder-less, roads that I had imagined were fine thoroughfares for the cowboys on horseback who carved them first as trails. The land was transitioning again from rocky deserts to smooth fertile plains. And although I had been distracted by the five hundred years or so of rich Hispanic antiquity in the regions old brick buildings, now I was on the hunt for a ghost I had been chasing my entire life.

The two of us had crossed paths on this Leg already in Lincoln, Silver City, Santa Fe and a handful of other locations before I made my way to the foot of his grave that day. Focusing on my balance as I stood in the stiff wind, I looked around thinking that the mud walls surrounding the tiny unkempt necropolis were quite modest for a man with a big name. Just happy to be there after a long stretch though, I conceded, "Appropriate I guess, given that he was an elusive coyote in life."

Digging deep into my pocket, I pulled the customary offering of a dime out, examined it by rubbing it a few times, and threw the coin towards his headstone to pay respects. I read aloud the inscription "PALS" etched there. Kneeling down, I started a conversation by introducing myself, then flattered with, "Well...you certainly left your mark young lad." A pause ensued and I asked, "Which alias would you give me if we were to meet?

Would it be Henry Antrim...William McCarty...William H. Bonney....or The Outlaw Billy the Kid?"

He still prefers the latter I'm sure, and I know it was a one way conversation, but still it felt like a meaningful one and remains a highlight of the trip. Growing up, and to this day, I've always had a fascination with "the Kid" and don't know why. Perhaps it was his adventurous side, the way my uncles referenced his name as they practiced games of quick draw, or that he was a Robin Hood type in a land where corrupt officials reigned.

His character I suspect would have been complex to say the least. Not a natural born cowboy of the West, he was in fact born in the sprawling metropolis of New York City, circa 1859. Known for his boyish wit, affinity for pranks, and loud contagious laugh, today he would have been that charming, cool guy that everyone likes. Even the reporters who painted him in a dim light to sell newspapers thought him to be highly approachable, courteous, and with articulate speech and written word, a pleasure to talk to.

Before you could see all those admirable traits, you first had to be fed the spite that the orphan Billy was a murdering bastard. A cattle rustler yes, but only after being forced into a fugitive lifestyle via a jailbreak that saw him serving a sentence for the crime of stealing. The tragedy of history is that he started out as a scared, hungry kid, but ended up public enemy number one. So it seems his soul was destined for an early grave and the conditions of the Old West simply facilitated it.

He was athletic and so had a reputation for speed with a gun. That skill preceded him and by the tender of age seventeen, already he was the target of every bounty hunter who wanted his name. Such was the case when gunfighter, "Texas Red" Joe Grant, tried unsuccessfully to shoot Billy in the back as the Kid walked away from a fight he did not pick. Hearing Joe draw, Billy turned to face his tormentor, drew his own pistol and from across the room of the saloon, planted a round right in Joe's chin, killing him instantly. After the duel, Billy said with humility, "It was a game for two, I got there first."

Good at what he did, that near same scenario played out a total of twenty one times to villainize him to many, but to glorify him to others. The fact that the poor of the region sheltered him in the face of large rewards was proof that they thought he was a hero. And so it was no surprise that the people of New Mexico mourned in July of 1881 when news broke that Billy, just 21, had been ambushed and killed by his old buddy Pat Garrett.

Standing at his grave now, I felt an abundance of loneliness there, not despair, just the tragic sense of loss that lingers when someone leaves us early. It's cold comfort to know that when people die young though, they never age, nor do their memories fade, instead they remain brio in our imaginations, and like Billy, their legends grow. Being only six feet from where he lay, I thanked his rebellious soul for all the romantic folklore he bore, then bid adieu and headed along my way.

Moving east I could feel my ears pop as we descended the high altitude plateaus of the West in favour of the constantly warming Gulf air. It was around day fifty of the trip now, and with my windows down, my now long shaggy hair tossed uncontrollably in the breeze. To my front lay the entire eastern half of the continent, and reflecting on the rewards of legs done, I wondered also what was in store. I reckoned though that if the road ahead was half as rich as the one behind, then it was gonna be a damn good time.

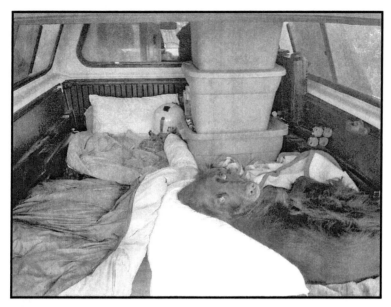

The back bed of my Toyota Tacoma made for an ideal home for the three and a half months Athena and I were on the road.

Stoney and I catching up in Edmonton.

Athena and I covered in frost after a long run in Northern Alberta.

Surfing for the first time in Tofino, British Columbia.

I left Oregon with a passion for barefoot running on the Beach.

Exploring the cliffs of California's serpentine Pacific Coast Highway.

Ironman Rich and I enjoy a couple of Irish Coffees (aka warm whiskey) at Buena Vista Cafe after a long run through the Redwood Forrest.

Stash and I silhouetted against the backdrop of San Francisco Bay at dusk before we part ways.

The Grand Canyon was just one of many iconic natural features that defined the seductive landscapes of the Southwest Desert.

Busted! Sharing a laugh with an Officer of the Arizona Highway Patrol after he pulled me over for speeding on an abandoned stretch of road in the middle of nowhere.

Exploring Antelope Canyon in Arizona.

Waking on Christmas morning to the vast red expanse of Monument Valley was an awe inspiring gift.

Visiting frontier ghost towns like Tombstone became the norm as I zig-zagged through the Four Corners Region.

Hiking trails in Utah's Zion National Park were lined by cliffs and waterfalls.

In New Mexico I enjoyed chasing the ghosts of Wild West Outlaws like 'Billy the Kid' as I visited the places they lived, died and are buried today.

Recreating Continental history was my only real road map and throughout the journey I was drawn to sites of significance like this one in Dallas Texas where JFK was assassinated.

Strolling the above ground graves in Saint Louis No.1 Cemetery in New Orleans Louisiana was interesting after New Year's celebrations in the French Quarter.

Leg 8
Texas The Lone Star State

"In the south there are no strangers, just friends we haven't met yet." – Southern Hospitality

It was rush hour in downtown Dallas as I dashed in and out of busy traffic to the beat of angry horns in a deadly game of Frogger. I had just spent several minutes on the sidewalk trying to time the daring dash, but caught in the fray now, and with cars whizzing by, I abandoned my plans and focused instead on getting the job done. Finally with a gap between cars, I lunged to where "X" marks the spot in the middle of the eastbound lane, squared my feet to it, glanced to the sixth floor of the Texas Book Depository Building behind me, then with a solemn shiver, snapped a selfie of a place I knew still harboured deep emotional scars.

After bolting back to the sidewalk, I turned and breathed a sigh of relief in the direction of applause coming from a local vendor who was selling souvenir newspapers of that fateful day. "See, that wasn't so bad," he said, laughing as he pointed out other tourists attempting the same. Approaching, I handed him five bucks for one of his articles and played it cool as I caught my breath, then patted him on the shoulder to say thanks. After reading the headline, I motioned to the surrounding area and said, "So, this is Dealey Plaza?"

His response was matter of fact. "Yup…this is the exact spot where John F. Kennedy was shot dead."

Being an admirer of the man, I had always wanted to visit the place where his life was cut short. Now there, I was surprised at how serene it felt. It was quite beautiful really, and I was struck that it seemed like an unlikely spot for a despicable act, but concluded also that assassins don't care for etiquette.

Born in 1917 into a wealthy Massachusetts family, JFK, for the most part, lived a life of frat boy privilege. Despite the silver spoon though, he was ambitious and seemed destined to rise to greatness. A man of service, he was first tested on the Pacific Battlefields of the Second World War, then put through the grinder of the House and Senate, before finally, and at only age 43, was elected by the people as the youngest leader of the Free World.

His mission during his reign was clear, to strengthen the country's defences, quell its ignorance, and advance the cause of Civil Rights while challenging tyrants abroad. We can say with certainty that he did all of that, plus much more in his short tenure. As with so many other historical greats I found myself chasing on this trip, his gift was a legacy of change that came at the cost of his life.

Looking around now as I watched others attempt to connect to his soul with the same foolish act I had just done, I saw that his adoring fans still flock here in the same way they did fifty years ago when he rolled through. The silent 8 mm Zapruder film of his assassination is proof of that. Taken close and from the angle of the grassy knoll, near where I stood now, it tells a story, but without sounds, smells, or the energy of chaos to put me right there, it also leaves a lot to the imagination. As a student of history, I can't help but attempt to recreate events to help fill those gaps.

Standing at the scene of the crime with the souvenir newspaper in one hand and my cell phone dialed to information of the conspiracy in the other, I pivoted in all directions to do just that. Soon I began to journey back in time. The paper was dated November 22, 1963, and I could see from the photos that little has changed since that day, except that the trees are much taller of course. To add to the likeness, it was a crisp fall afternoon under the same Texas blue sky that will never change.

To my left on the Grassy Knoll is Mr. Zapruder with his state of the art colour motion picture camera rolling. To my right crowds of families with young children gather and play at the park opposite the red bricked depository building, they are hoping to get a glimpse of the President as he passes by. It's not long before heightened commotion infiltrates the bliss as spectators stop to stare, and whispers abound that a shiny Lincoln Continental Convertible has been seen moving along Houston Street towards Dealey Plaza. Those rumors turn to cheers, and whistles of celebration to direct all eyes to the intersection ahead where I expect to see a caravan emerge any minute.

Straining to peer over the crowd, I stand on my tippy-toes and can see the spinning lights of his police motorcycle escort round the corner first. Seconds later I can make out two small American flags fastened to the hood of the lead car as it emerges. Smiling, I realize it's the mark of the President himself, followed as always by the entourage of a security detail. All vehicles turn left onto Eastbound Elm and head straight towards me.

Only thirty yards out now and my heart skips a beat at the clarity of what I see in real time versus the fuzzy images the world has come to know. Sitting beside the President is his loving wife Jackie, together they are the portrait of Americana. The sun's glare is bouncing off the black asphalt to blind them both, and to combat the annoyance, the President squints and sweeps with one hand, his feathered hair made slightly array by the gentle breeze. They are both smiling and waving to the crowd affectionately.

Through the noise of support, a slight pop rings out to jolt the President, it could have been anything but heinous in my mind, but instinctively I look to my left to investigate. Seeing nothing, I glanced back to his car to realize his head is slightly forward of uncomfortable now, and his hand is shielding his neck. Still, not a single sign of distress comes from Kennedy as he begins to pat his own body down for the source of a stinging sensation.

Ten yards out now and I observe Jackie's jubilance turn to intuition as she catches her husband's unusual behavior in her peripheral and leans in to check on his welfare. Her movements are made obvious by the contrast of her stunning pink suit against the drab black car and so people take notice. He looks conscious but doesn't respond, and Jackie, still unaware that a bullet has just ripped and ricocheted through her husband's body, is startled as she realizes blood has begun to trickle from his mouth.

Instinctively she gazes into his eyes for answers and a second shot rings out, this one though is loud and piercing enough to make me jump and send the rest of the spectators diving for cover. Confusion ensues as the President's head snaps forward, then backward, then scatters his brain matter. Horrified bystanders wonder if they have just witnessed another of America's frequent assassinations.

The smell of burnt gunpowder makes it clear they have, and most stay glued to the knoll waiting for another salvo, but nothing comes. Reverberating through the plaza however is the sound of squealing tires and Jackie's wailing voice as she reaches towards the Secret Service Agent being dragged by the limousine, not to help him, but to gather the fragments of her man's skull.

Coming to reality as a town car sped past, I absorbed the surroundings for another minute or so while attempting to validate the conspiracy theories that have formed over the years. The talk has always been of an inside job, but because the offender was caught, I've always felt otherwise. And as a retired army sniper with experience enough to say that, I couldn't help but scrutinize the scene to give credence to my thoughts.

Examining the area, I acknowledge that the plaza is a hunters dream, but not a sniper's paradise as the conspiracists would have you believe. I can say that with confidence because even with considerations of slant angle, a moving target and obscuring foliage, it appeared that from all directions the flight time of the round would have been supersonic short. Meaning it was an easy shot, and because it didn't have a chance to tumble with instability, it didn't take much skill to execute. To take credit away from a crazy man, a child could have hit more targets of less opportunity than Oswald did. He was a sharpshooter, not a sniper, there is a difference; all snipers are real good shots, but real good shots don't make snipers.

The trade is a distinct qualification within the military, and not one handed out lightly given the resources, time and money it takes to produce one. Just to be considered for pre-employment training, soldiers must first volunteer for an open competition, get physical and psychological clearance, then interview with the Unit Master Sniper and his staff to ensure they're fit for the culture. Once that's done, minimum prerequisite are reviewed to verify competitiveness with attention given to candidates who maintain top fitness and job assessments, have a set number of years in rank, and who carry assets like reconnaissance or other specialized qualifications. The dues paid on a couple of overseas deployments doesn't hurt the resume either.

Jumping through those hoops to make selection though is the easy part, because if you are offered a spot on course, ahead of you is one of the most grueling programs you'll ever challenge. Designed for maximum attrition, it's administered by hardened instructors who are never impressed, have been there and done that, and who would rather send students packing then invite them into their covert world.

My pre course started with sixty high achievers, three months later we limped into camp broken, battered and bruised by the curriculum, but with six newly minted Canadian snipers to send out into the field. It was a tough course, and to this day learning I had enough points to graduate with my peers and receive my silver coin to prove I did remains my proudest moment, and the coin, my most prized possession.

It is tough for a reason though. The title comes with a lot of responsibility, shooting being the least of what's expected of a shooter. Considered "force multipliers" they're trusted alone or in pairs, and have the primary role of observation and intelligence gathering. But when called into action, their real meat and potatoes is earned by tracking, stalking and taking out targets with the "one shot, one kill" reputation their known for. These guys don't come home until they get the job done.

Guardians of the troops they are tasked to protect, they wear no rank or identification of any kind, and save for the odd rumour that "Call sign 66" is working the area, nobody but their commanders knows where they are. The clue is given when the mission is deemed fulfilled, and over the crackle of the radio comes the words the shooter has been waiting for, "thank-you '66'... go ahead and reach and touch 'em." The order doesn't even finish crossing the Battle Group's Net before a single crack of his high caliber rifle can be heard echoing across the valley to silent all within earshot, who wonder where in plain sight is he hiding.

Leaving Dealey Plaza I was certain the conspiracy question was put to rest in my mind as nothing more than the opportune work of a patsy looking for a place in the annals of history. Had it been a trained government sniper, as theorist's suggest, the shot would have been much more complex and with untraceable depth. Trust me when I say, professional shooters don't get caught.

※ ※ ※

On my third day of exploring the great State of Texas, I was en route to Houston, but in a roundabout way, that is to say I was backcountry 'road'n' it all the way. As I meandered through rural communities I couldn't shake the welcome sense I got there, or the breathtaking scenery that awaited me. It was golden plains and farming operations mostly since about Lubbock's. Some had freshly turned soil in anticipation for spring, while others were still buried by winters flattened crops. Nearly all though were decorated in some way by a service rig, combine or coulee.

To add a bit of charm to the obvious was a taste of good old fashioned politeness. Everywhere I went people with names like

"Jimbo" or "Barbara" or "Bubba", greeted me with "Sir" and offered hellos while holding doors. I recall during a coffee break at a service station outside of a place called "Gun Barrel City," that I embraced those values and tried my best to return the favour.

I was people watching from Taco's tailgate when a family in a mudded pickup pulled in. Dad, who wore a cowboy hat and a clever belt buckle, was driving of course, and beside him Mom sat scootched up tight, while in the passenger seat sat a dog who owned shotgun. The kids were in the back box of the truck and were just in the process of bailing out. As Dad walked by, he gave me a nod and said "howdy." I nodded back as if we were friends and he waved before he went in. Those bits of simple courtesy happened over and over in Texas, and it's safe to say I felt at home in the "The Lone Star State."

Nearing the coast, I was happy to be lured by salt air coming off the gulf and the knowledge that soon I would be arriving at the iconic centre of America's oil wealth, which is also known as "Space City" USA. Did I mention I'm a bit of a Cosmos enthusiast yet? If not, be warned that my dream job was to be an astronaut, I've since settled for stargazing on hilltops.

Once in town I headed straight towards the Space Center for a day of touring that I wouldn't soon forget. Entering I was greeted, fittingly enough, by the dreaming works of Leonardo Da Vinci's machines in motion exhibit. It was interesting to see that his creations could hold their own against the titans he inspired like the shuttle, Lander and Space Station displays. I saw firsthand that he was a man of enterprise.

Strolling past those and to a collection of space suits, I stopped to take selfies, then ventured into a dark room where pictures hung that I recognized. I smiled because they were of the USS Hornet recovering the same Apollo capsule Stash and I had explored in San Francisco. Following the same dimly lit expanse, I came across, to my surprise, an excess of celestial objects in a thick glass box. They were the crown jewels of the Universe and looking at them, I whispered, "Moon Rocks." They were rubbed as black and shiny as Hematite, and I imagined they contained the energy of the Sea of Tranquility, a feature of the lunar surface I had observed a thousand times before from a telescope in my own backyard while smashin' back beers.

It was a unique connection, in all I spent about six hours there discovering things I didn't know. As I prepared to leave, I passed by another national treasure and stopped for a picture. It was the actual Kennedy podium from which he gave his famous space address speech. Looking in every direction to make sure I was alone, and feeling like old JFK and I were buds now, I jumped the rope barrier, put my hands where he would have, then attempted my best New England Brogue by reciting a few words, *"Ask not what your country can do for you…but what you can do for your country."*

I began to chuckle that the accent was probably highly offensive to anyone from Boston, then noticed my audience wasn't laughing. There at the door was a very displeased elderly security guard giving me the angry eye. I gathered from his reception that he was used to dealing with tourists doing the same, and instead of questioning my motive, sternly he pointed to the door and said, "LEAVE NOW."

"*I know when I'm not wanted,*" I thought with a huff before walking towards the parking lot doors. It was a bit embarrassing though that he escorted me out, still I didn't have the heart to tell him I was leaving anyways. Oh, well.

It was dark and raining as I drove towards Louisiana, blinded at times by the incessant flashing of high beams from oncoming traffic. Wondering why people were flashing so obnoxiously, I returned the favour, but with double taps, and the odd gesture too, not that they could see that. In fact I was so engaged in the game that I totally failed to realize I was being tailed again. That is of course until the ghost car lit me up.

"*Oh c'mon…*" I thought as I sighed trying to count the number of times I had been pulled over of late. A few minutes passed before I noticed the swaying motion of the officer's flashlight coming towards me. Tapping on the window with its blunt end, I rolled it down and was greeted by its business side. The cops inadvertent flip of his torch blinded me for a moment and to sell it as annoying, I forced an aggravated look.

Seeing my reaction he canted the light away so it wasn't directly in my eyes and said "Oops, sorry about that…are you okay?"

A tad taken aback by what seemed like an apology, I responded, "Ummm lovely…and you?" He lowered the flashlight completely now, then leaned forward to poke his face inside of Taco. I could tell he was searching for signs of alcohol, so I breathed heavily to prove I hadn't been drinking at all. Pulling back, he said, "I'm fine and thanks for asking," then paused to look over his shoulder at the speeding traffic and finished with, "Well, I don't know about folks up in Alberta, but in Texas we use headlights in the rain at night."

I shot my left hand to the light switch and realized it was in fact in the off position, then smirked as the pieces of the puzzle came together on why so many had been flashing. Based on his comment, I presumed that he had already run my foreign plate, and so I looked up to address what I thought he was gonna say.

His silhouette from my angle painted a tall, clean cut man with a laugh as jolly as Friar Tuck's and a belly to match. I felt relaxed by his easy presence, and before I could talk, he cut me off with a snicker and said, "So unless you're a Canadian bank robber, which I highly doubt, you might wanna do the same before I get a call that someone's been killed."

Flushed with embarrassment, I shook my head to apologize. But as if he knew the reason why, he continued, "I know, I know…I own a Taco myself so I feel your pain." He was referring to the fact that a Tacoma's interior gauges illuminate when the key is turned, verses like most vehicles, when the headlights are switched on. "Just be cautious from now on," he reinforced, then asked, "Are you working oil down here?"

"Nope, just passing through," I replied with a short answer, very aware that he was in the rain and I wasn't.

"Ya it's a nice couple of days drive to Alberta...I do it annually in the fall to go big game hunting in the mountains of the Crowsnest Pass...do you know the area?" His words radiated endearment and brought a prideful twinkle to my face as I pictured in my mind the autumn foliage of a million trembling aspens all quivering at once on the slopes of a place I know and love well.

"Kinda...real big area though..." I responded with a nod so as not to interrupt.

"Ya, it's Heaven up there," he said.

Next he changed gears by digging for details in a noninvasive way. "Did you hit the storm coming south on I-25 through Wyoming? News said the Interstate was still closed..."

I still wanted to keep my answers short and sweet so he could get back to the warmth of his cruiser, but had to answer truthfully despite knowing it would prolong his exposure. "Actually, I took the back roads the whole way."

The admission seemed to raise a few flags, and I could tell his cop senses were tingling now. Canting his head to side, he leaned in and placed both hands, plus his weight, against the door so I couldn't open it, then said, "Now why would you do that? It would have taken a week...maybe more?"

I sensed the tides of camaraderie were turning, and wondered if he thought I was using the back roads because I was up to no good. I now jumped into the conversation full bore. With a smile of my own, I offered, "More like six weeks. I'm on a North American road trip."

A short pause followed as he considered what I said, then he blurted out, "That's cool!" His enthusiasm let the 'Friar' out of the bottle again. Happy to be on the same page once more, we began to talk about our mutual divorces and small business experience. I was a little surprised though that he didn't inquire further about the trip. Instead his interest, once he found out I was a landscaper, lied in questions about paving stones and rock beds.

After a brief exchange on the subject, that saw him kneeling to the ground on one knee to map out what he had installed in his own yard, which was strange to see a lawman do, I looked to the traffic piling up and felt like asking "Ok, am I on candid camera here?" He must have noticed that I looked concerned, because almost instantly he honed in on the traffic too, stood up, straightened his uniform, then said "Man, I gotta get out of this rain," then wished me a safe trip and moseyed back to his patrol car.

I looked to Athena dumbstruck, "That's weird...he never even checked my licence?" I conceded his friendliness did fit the bill of Texas. Driving away I grinned. "That's gotta be either the nicest, or the loneliest cop I've ever met!" Then noticed in my rear view mirror that he was flashing his lights furiously. Reaching down to check my own, I realized they were switched off again and turned them on. *"Murph"* I thought, certain that the Friar was having a good laugh. He made my night, and one thing is certain, in Texas, hospitality isn't forgotten.

Leg 9
New Years In The Big Easy

"There are far, far better things ahead than any we leave behind." - C.S. Lewis

It was noon on New Year's Day as I wandered aimlessly past the grand buildings of Bourbon Street in the Old French Quarter. But everywhere I looked there was proof that the party was still going from the night before as groggy people danced to the sound of Jazz music filling the air. To add to the atmosphere, literally, was the scent of oil and batter from food vendors who pushed a mirepoix of sweet, sickening, deep fried Creole cuisine. Normally such fair would be appealing to me, but suffering the effects of dehydration, a splitting headache and the hangover from Hell, the smells did more to convince me of just one thing; I had to puke, and I had to do it right now!

At first I fought hard against the sensation, then realizing I had no choice in the matter, I swallowed my pride and bolted with both hands covering my mouth to the refuge of a back alley. Once there, I leaned forward and forced a heave to induce a salvo of projectile vomit that left me relieved, but exhausted and gasping for breath. With the sense that every vein in my face was now popped, and that everybody had watched me chuck, I wiped the sweat from my brow, gathered myself, then did the walk of shame back out passed a family with kids, thinking as I did, *"Ohhh that was a nice show, Aaron..."*

It didn't really matter though, because now I felt a million times better, and as I walked, I could feel the pep in my step return. Passing a pair of buskers that I recognized from hours earlier, I threw a couple of bucks their way, then forced a path through a crowd of tourists gaggling nearby. Once on the opposite side of them, I emerged to the step of the same twenty four hour pub that was partially responsible for my misery now, and headed inside.

Entering with bloodshot eyes, uncombed hair, and breath reeking of Jack Daniels Whiskey, I knew I looked to be in a sorry state. Beyond caring about self-awareness, I went straight to my old stool at the end of the counter and sat down. There I slumped for a minute, then tossed my last crumpled twenty on the bar and signaled to the barkeep for a pint of beer. As he poured the dense frothy draft from the tap, my mouth began to water at what I knew was coming next. He was a pro, and after knocking the excess foam off, he slid the brew to within four inches of my open hand. "Cheers", he said, before walking away.

"Cheers indeed," I thought, as I raised the glass, then tipped it in his direction to give thanks. Slamming the brood back in a couple of gulps, I brought the glass down hard, but playfully, and with a monstrous burp said through a grin, "One more of the hair of the dog that bit me please." The bartender turned and smirked at a line he knew well, palmed the twenty, poured another, then passed it over, but this time with a bottle of Advil, and said prophetically, "Welcome to the Big Easy."

It was a New Orleans baptism by fire, and although usually I don't drink so heavily, somehow the night had gotten away on me. Grinning as I sifted through blurry pictures captured on my phone, I realized I had been on some sort of solo parish pub crawl. Each image told a story of a different location, but all were filled with the same narrative of smiling faces from people I didn't know, and had stacks of empty shot glasses scattered about. Looking at them, I was certain of just one thing; it was a hell of a good time last night. Now if only I could remember what I did.

Founded by France in the early 1700's as another of the Kingdom's strategic fortresses along the Mississippi River, today New Orleans is a modern melting pot of those same influences, but mixed of course with the skyscrapers of Americana. I was thrilled to be in one of the oldest cities of the New World and as I walked its cobblestone streets trying to leave the hangover behind me, I reveled in the chaotic architecture of the French Quarter.

With colonial clout everywhere, and countless layers of nuance to prove it, the place was charming, mysterious and spoke with incredible character. Structures ranged in size. There were small wooden townhouses colourfully painted in pastel tones that complimented the bright, lush, flora of their manicured grounds. Others were large multi-story brick buildings, like the behemoths of Bourbon Street that dominated with earth tones, wrought iron balconies, and walls covered in lavender vines that clung to cement mortar spilling from joints senior even to the Civil War.

Pushing the leaves aside to admire the masonry beneath, I beamed that each brick lay twisted in an unplumbed fashion as perfectly imperfect as the city itself. That they've stood the test of time is a fine tribute to the tradespeople who built this town, and wanting to connect to their roots, I did what I often do when I'm in a new place, I headed to the local cemetery. Its sounds strange I know, but one of my favourite things to do in a new city is to visit its graveyards. I find their spaces peaceful, relaxing, intriguing and not at all unsettling. They remind me that no matter how much we have in life, in death, we all share the same fate, and there is comfort in that, for the little guy anyways.

I walked to Saint Louis No.1 Cemetery on the edge of the Quarter. Small and walled, it spans just one city block, but is the final resting place to many thousands. Strolling through, there was a Gothic feel to the place, and I recall at times having difficulty squeezing through some of the narrow gaps between the above ground graves. Still, I enjoyed running my hands over their moss covered lids pocked by the stamp of time, and I couldn't help but wonder who might lie inside.

Some of the tombs housed the bodies of the tradespeople, whose work I had just admired, others were of prominent leaders, politicians, and even Voodoo doctors. While still others were empty, but in the queue for future residents like actor Nicolas Cage who will one day call Saint Louis his final resting place. The difference in size and condition of crypts varied greatly. Most were humble, crumbling, but to me, nice little abodes not much bigger than a typical casket. Others however, were grand shrines that could rival a house. And although many of the larger structures were built for entire families, which was a tender notion, some contained the body of just one individual.

The contrast seemed demeaning at first, but paralleled society to a tee. And I pondered whether it is the force of vanity, insecurity, or fear of death that compels the rich to advertise their prestige long after their demise. Whatever the reason I'm sure they'll be humbled when they learn the value of a soul has nothing to do with a bankroll.

In fact when I meet my maker, I hope my family wastes not a single dollar on a box of any kind, but instead dresses me in my army uniform with flip flops on, then wraps me in a flag and buries me under an unmarked larch tree overlooking the Alberta Rockies. That way while the tree grows, I can forever embrace the sunlight on my face as I watch the season's change on the same eastern slopes that captured me in life. Now that's my idea of eternity.

After leaving the cemetery, I left New Orleans too, and spent the next couple of days working the small roads and trails of the Creole State. The landscape was flat now, and dominated with coastal marshlands, bayous and inland forests that were filled with towering Louisiana cypress trees. Since I left Calgary, the road had taken us from sea, to sky, and now back to sea again, but previously was always paved over solid ground. Now as I negotiated these new massive watersheds, it was apparent that ingenuity played a big part.

The stretch had given way to a stitched complex system of bridges and causeways that hovered effortlessly over great distances and atop thousands of concrete piles drilled deep into the swamps. It was a perplexing environment, and although I wanted to get lost in it all, I wasn't naive to the fact that as a prairie guy, I was literally out of my element. Feeling vulnerable to the snakes, alligators and big cats that call the area home, I kept a bit of distance from the wilds. I did venture in a few times, but only for short runs, and when I did, I stuck to roads instead of the trails I usually set out to find.

Meandering along, I made my way into Mississippi, then Alabama as I searched for hidden gems. There I fell in love with the story of its turbulent, but resilient beginnings. For some time I had wanted to explore the Civil Rights Trail in the region, but was disappointed that when I arrived, a storm was rolling through, so I wouldn't get the chance. Strong winds, rain, and even snow, saw that most businesses were closed and that I was literally left out in the cold.

Because of that, I had scarcely left the comfort of Taco's warm cab and was starting to succumb to the effects of cabin fever. Wanting to stretch my legs out one last time before I left the Deep South, I resolved, that despite the sheets of rain, I needed to go for a long run. It's never fun to grind miles out in the stinging rain, but if Forrest Gump, the State's most famous son, could do it, then so could I.

Parking just outside of Mobile, I changed quickly, then headed out into the pouring rain. Within minutes my fingers were numb and I could feel water sloshing through my toes. Distracted by the elegance though, I followed the road left, right, then left again while attempting to get lost, and eventually I did end up on an old spooky rural route that looked like the South I had seen in books.

"This was the Confederacy," I thought, of the beautiful countryside while stopping to admire it all. The expanse was littered with old mansions whose property lines were defined by rows of ancient Oaks, and I couldn't help but wonder if they once bound the cotton plantations of the Antebellum South. If so, that meant these living beings were time capsules of a sort, and that they also harboured crimes of the past in their roots. I began to pass under their limbs and was mesmerized by the Spanish moss that swayed and dipped eerily in the wind, then invited me over. Once next to one, I leaned in, then asked, "what secrets do you possess my friend?"

The question, combined with the sound of pounding rain, conjured an image in my head that disturbed me greatly. In it I could hear the cadence of steel shackles being drug by the bloodied ankles and barren feet of the damned. Preceding this parade of indignity was the voice of an insulting slave master riding high on his horse as his villainous dogs barked threats and drooled at people who never did anything wrong. To entice the beasts, their bodies would have smelled foul from sweat drenched rags made tattered and pierced by a lifetime of working in the sun.

It was an abhorrent visual, and my stomach turned to knots at the thought that humanity during that period was lost. Being a Canadian I can't possibly understand the politics of racial division down South. But I don't need to, to know that "all men are created equal" by the hand of God, so there could be no justifying it as right. I cringed knowing that such injustices took place where I stood, then conceded that America still has a long road ahead to reconciliation.

Getting back to Taco after the run I was shivering and felt sluggish so cranked the heat on. With the windows fogging up, I struggled, as I always do, to change from running shorts into jeans while hovering over the passenger seat. It's never a graceful task, but on this day it felt a little more arduous. The jeans were tight and I wondered if it was because I was wet, or if maybe they had shrunk. "Wishful thinking!" I joked aloud with a chuckle before patting my belly and admitting to Athena, "I think I grew an inch or two…"

True, I had taken a couple of days off running because of the storms, but I knew that wasn't the reason my jeans were bursting at the seams. A line from a "Simpsons" episode explained it best; in it Lisa employs flattery on Homer by suggesting he's somehow miraculously getting stronger. His reply was a sheepish one, but predictably masculine, "Well...I have been eating more," he says as he flexes his muscles.

And like Homer, I had too. Since I arrived in the South, I had picked up a bad habit, the all you can eat buffet. Typically I'm a light eater, and so far on the trip had enjoyed a healthy diet of convenience foods in the form of raw fruits, nuts and plenty of water to keep fit. But lacking energy of late, and sporting pants that revealed all, I realized I needed to get off the gravy train right away. But as with any vice in life, I figured "one more time wouldn't hurt a thing."

The problem wasn't the buffets, it was that my eyes were bigger than my stomach, and so I would fill a tray packed with chicken, beef, pork, potatoes, dessert and whatever else I thought I could cram down my gullet. Then because I can't bring myself to waste food, I would spend the rest of the night getting 'food drunk' trying to finish it. Do that for a bit, and it's not long before the pounds start to pile on.

Growing up with a single mother who fed three kids while living hand to mouth, I remember the disappointment of a fridge with nothing but a bottle of ketchup in it. The feel of hunger pain was real back then, and is still hard to forget now. So to this day when I see food, I don't just see a meal to be taken for granted, I see finite energy harnessed from the sun, earth and water. Then on top of that, there's a complex chain required to get that harvest, and it makes me realize that eating is a privilege, not a convenience. The bottom line is that with so many people starving in the world, food is a gift we can't in good conscience waste.

As I sat in the dining area reminding myself of that, I couldn't help but eavesdrop on a conversation unfolding in front of me. A couple was arguing about something at a table adjacent, and I wouldn't have paid much attention, but the look of their children's faces gave it away. They seemed horrified as the fight escalated to include threats and foul language. Finally, and to the relief of everyone, the restaurant's manager approached to ask the couple to lower their voices. I'm sure they didn't even realize they had an audience, because when the dust settled, they looked embarrassed. The signs of their divorce were evident, and resuming my meal, I sighed, "Been there."

I'm a traditionalist when it comes to marriage and the roles within. So naturally I have this romantic notion that the place of a mate is reserved for just one. I know also that if you find that special someone, hold onto to them, because they're as rare as diamonds. A life partner is more than just somebody to share space and debt with. They should, without condition, be your crutch and the one you're most comfortable with. If that's found, then the union will be full of intimacy, adventure, adversity, impulsiveness, routine, and the kind of rewards that transcend illness, and time, until that final day of reckoning arrives, when you can both look into each other's eyes and say with conviction, "I was your biggest fan."

I believe that, but like most I've also been duped a few times, and so have learned what the couple who seemed beyond help were figuring out. That is when the inside jokes, laughter, and simple pleasures that ignited the flame are replaced by the noise of resentment, spite, and ignoring nights, then the magic is gone and it's time to move on. I paid my bill at the register at the same time they did and smiled in their direction to let them know we're all human, and that the future is always brighter than the past.

Driving east out of Alabama and into the Florida Panhandle, the rain began to relent and as it did I reflected on a good stretch of road over the last couple of days. I was feeling a bit antsy from being cooped up in Taco though, and knew I needed a jolt of adrenaline to spark my adventurous side. A challenge was brewing in my mind and butterflies stirred as I considered whether or not I had the parts to follow through on it.

Leg 10
Florida Sky's Away!

"Once we accept our limits, we go beyond them."
- Albert Einstein

With hands shaking and sweat beading across my brow, I pulled the fogged goggles from my face to wipe the condensation off, hoping, unsuccessfully, that the act would distract me from the fact that the pilot had just turned and yelled, "ONE MINUTE!" Rubbernecked and wide eyed, I nodded to acknowledge him, then played it cool as I searched subtly for a way to tighten the clips of a harness that bound me to a nice young Brazilian man, and wondered as I did why nobody but me seemed to notice that the door of our plane, soaring at 18,000 feet, had just opened to reveal the vacuum of space.

If the Brazilian had noticed, he wasn't saying anything, and perhaps thinking he would spare me the anxiety of it, instead, he simply just smiled and pushed my body towards the exit. Resisting at first with dead weight, my reluctance cycled to acceptance, and aware that this stupid idea was my own, I joked, "WHAT WAS I THINKING?"

This was the morning of the day I decided to close the chapter on my Continental Space Odyssey with a bang, or a chute as it were. So far on the trip I had stumbled onto the Apollo in San Francisco, solved a long standing mystery of alien conspiracy in New Mexico, felt the energy of moon rocks at Mission Control, and even gave an impromptu speech from the President's own Space Race podium in Houston...well, I almost did.

Now as I sat with consternation and about to embark on the world's highest tandem skydive over Cape Canaveral, the launch site of NASA's shuttle program, it occurred that men like me aren't meant to soar where astronauts go, because as you know, we're afraid of heights.

The thought provoked a fit of nervous laughter, and I shook my head in knowing I'm neither an intellect, nor a philosopher, but rather just a guy who lives by the creed of "man-up." With that promoting courage, I evoked another card to solidify my resolve and repeated aloud, "It's just ten minutes for the rest of your life…"

Once at the door, I swung legs out and shimmied carefully along the roll bar. If I thought I was nervous before I stepped foot outside of a moving aircraft, I would be remiss, because it felt like things just got real. Out there the roar of the wind and engine was deafening and enough to take my breath away. With a big gulp to draw in as much oxygen as I could and muster strength, I gave thumbs up to my Brazilian counterpart, then crossed my arms and waited for what lie ahead.

He began to rock us back and forth on a ledge at the edge of the atmosphere. The first two teeters we gained momentum enough to bring my gaze forward and onto the width of Florida's Panhandle, then aft till they were lost in the realm of the Cosmos. On the third rock, the Brazilian Man gave one big thrust to push us past the point of no return, and as he did, we rolled from the platform to let gravity take us to Earth.

Flung to fate, and tumbling like wet shoes in a dryer, I struggled to get my bearings as we spun heavy G's in a furious motion that flailed my limbs and divulged only the horizon, the blue ocean, and the glare of yellow sun, over and over and again, until my faculties were sickened by dizziness not felt since childhood days spent on the merry-go-round.

It felt chaotic and I wondered if all was okay, then a tap on the shoulder advised, to my relief, that I wasn't flying solo. After signaling me to spread my arms like wings, as we had rehearsed, we began to stabilize with our bellies pointing down. Soon my angst was gone, replaced instead by felicity, and for the first time I was able to grasp with clarity the freedom of the immense blue yonder.

Descending at a rate of two hundred feet per second towards a mountain of fluffy white cloud that looked like a solid structure from above, being pierced by spectrums of orange light not seen from below, I instinctively braced myself, then radiated as we passed straight through and into an empire of blinding fog reserved for the gods.

Emerging from its flat bottom, my ears popped relentlessly and I noticed the air was tropical warm again, and strong enough to cast waves like tossed linen across my skin. Looking past my hands, I could see the launch pad below, and as I readied myself for the next cloud I wondered how far we had fallen. The freefall lasted the better part of two minutes.

If that doesn't sound physiologically taxing enough to get the adrenaline pumping, I invite you to attempt a quick exercise with me. Try if you can, to find a quiet room where you're able to stand motionless for two minutes. Once there it would be helpful to have someone keep track of time or set an alarm for you. Begin by lowering your head and gaze to the floor, then close your eyes. Now imagine you are standing on a steel bar beneath the belly of an airplane, in high winds, and miles above the ground where there is no way out of the predicament, except to jump.

When you're ready, signal for the clock to begin, then visualize yourself falling away from the bar until you're moving feet first and approaching the speed of sound. As you do, imagine the crushing effects of changing pressure on your overworked senses, as fear works to manifest the nightmare that has terrorized your dreams; you are plummeting to earth.

Watching the ground advance, your belly fills with the same butterflies that flutter with each drop of a roller coaster, but these ones don't subside and soon rumble ten times that to the point of pain. The splash of death seems imminent now, and as your conscience reconciles the fact, you claw at thin air and beg the question, "Why did I jump from that perfectly good aircraft?" If you can afford the time to do the exercise, you'll be surprised at how long it takes to kill a two minute freefall.

At around six thousand feet we deployed our shoot and sailed the rest of the way down with the grace of an albatross on calming, silent currents. Smiling after we landed, I was in a state of euphoria for some time as others mocked that my hoody and clothes were clinging unzipped and disheveled as if I had just stumbled onto a horde of meat hungry ladies at a bachelorette party.

I had taken the plunge before, but never ever from such intimidating heights. Buzzing with a sense of accomplishment from pushing the limit, I thanked the staff of the Skydive Space Center, then received my token certificate which reads, "Aaron has successfully completed the prescribed ground school, passed all practical tests and professionally executed the WORLD'S HIGHEST FREEFALL SKYDIVE from an altitude of 18,000 feet."

Afterwards I headed to the Kennedy Space center to finish the day with a tour of the NASA grounds. Walking past the Rocket Garden and into the main building, I was pleased to find I was just in time to have lunch with an Astronaut named Barbara Morgan. She was nice, and I told her about the jump and said, "we're pretty much equals now." She laughed hard until I left the room.

From there I was guided into a dim hall with curved walls and the sound of radio chatter to draw me in. Curious of it, I moved slowly through until I arrived at the end and was taken aback by what I saw. Suspended belly down, and on a slight angle with cargo doors open under a gauntlet of directional lights made to resemble sunlight piercing through the darkness, was a dreadnaught of exploration, the actual Space Shuttle Atlantis.

It was a charming discovery to make and I enjoyed strolling around her with the perspective of a moon walking cowboy in awe of the vehicle's size. With familiar white body, worn grey heat shield, distinct A-frame wings, rudder and giant engines, her run of presence was some six stories high. *"This was a workhorse,"* I thought of a machine whose run-of-the-mill exploits included pedestrian service to Russia's MIR, NASA's International Space Station, and the Hubble Telescope.

With a hundred and twenty six million miles logged in the process, or about five hundred and twenty five tickets to the moon, Atlantis certainly did her part for mankind. Moved by memories of seeing her blast off many times as a child, I said, "So nice to finally meet ya, old girl." I snapped a selfie to end my Space Odyssey.

*   *   *

I spent the next couple of days driving south by way of U.S. Route 1, a small historic secondary highway that skirts the East

Coast. At around twenty four hundred miles long, the road is America's longest north-south corridor, running basically from the Canadian Border with Maine all the way down to the tip of Florida, and it was one I would use often in the weeks to come.

Traversing through agricultural areas, the stretch connected pine and cypress forests to the palm-lined avenues of retirement communities, then to the sticks of the Everglades. I stopped in Vero Beach to explore those swamps, because Ironman Rich from San Francisco was kind enough to call ahead and set me up on a tour with his old pal Bill who owns an ecotourism business called "Gator Bait."

Arriving to the dock a couple of hours early gave me time to organize my pictures of the Deep South before Bill arrived. Once I saw his truck and branded airboat pull in, I waved and approached to introduce myself. Rich had warned that Bill was a nice a guy, and based on my initial impression of him, I couldn't agree more. He was dressed in his Florida best; a tan jacket, blue jeans and a ball cap that exposed skin as sun kissed as the Citrus Belt I had just passed.

I told him I was "Ricardo's" friend from Canada, then we both had a good laugh at a joke he made loosely at Rich's expense, not because it was terribly funny, but because the Ironman wasn't there to defend himself. With the ice broken now, he showed me his rig, broke down the activities for the day and welcomed me aboard.

The prop of the boat pushes air not water, so it creates little or no footprint in its wake and allows for an intimate close brush with nature. We began by crisscrossing six foot high rivers of saw grass that ran like roads over the flat landscape of wet prairie, marshes and sloughs dotted by colourful flocks of wading herons, egrets, ibises and pelicans.

A half hour into the tour, Bill stopped the engine to allow us to embrace the eerie quiet of a place that teemed with life. Directing my attention over the edge of the boat, he pointed to the water and said, "Come take a look." I peered into the water to see what he meant. There in the rusty shallows was a cornucopia of insects, fish and frogs scurrying about at the base of the reeds, I smiled to let him know I was impressed. Sensing I was lost, he pointed again with "No, not them...over there, do you see him?"

This time I leaned hard until my face was maybe twelve inches from the water. Focusing for movement, I saw nothing, until finally in the calm materialized the mask of a toothy grin which sported rows of pearly white teeth and a set of blinking black eyes. Startled, I stumbled back into the boat as Bill used laughter to break my fall. "Did you see the gator that time?" he said to mock me. Deflecting, I answered sarcastically, "Ohhh, was there a gator down there? I didn't notice…"

I had never been this close to one before, and feeling intrigued I leaned in for a second glance. He was somewhere right before my eyes, but inanimate and with monochromatic hide against the muddy bottom, I struggled to pinpoint him this time. Knowing that the chameleon could conceal all, except that predator's grin I saw, I searched for teeth again and soon was onto him. Tracing the silhouette of his body with my finger, a picture of a remarkable beast began to emerge. Heavily muscled and with a robust head, wide back and long dragon ridged tail, I joked that without Bill there, I would have tripped right over him. As we prepared to leave, I pointed into the water and said, "You, Sir, are *exactly* the reason I was so timid back in Louisiana."

The day was rewarding and after getting back to the dock, I thanked Bill and fired Rich off a text as well to do the same, then jumped behind the wheel of Taco to hit the road again. Continuing south, we stopped in Miami for a night, then at the crack of dawn, headed towards the coral cay archipelago of the Florida Keys via the Overseas Highway. I had no idea that Route 1 carried past Miami-Dade, so naturally when I stumbled onto the leg, I felt obliged to keep going.

The stretch is undivided, adorned by forty two bridges and spans roughly one hundred thirty miles down the Keys towards the Southern extent of Continental USA. Virtually unchanged for decades, it was a serene drive back in time. Wherever I stopped, everyone seemed to be flip flop friendly, and the place was more than postcard beauty, it was the image of eccentricity. Looking to my left I saw the expansive Atlantic Ocean, to my right, the Gulf of Mexico and everywhere in between the quaint Caribbean influenced communities. It felt welcoming there, and I could see how the lifestyle would have inspired great works by people like Tennessee Williams, Ralph Lauren and Ernest Hemingway.

The allure of the salt brine in the air was powerful, and wanting to experience the abyss, I stopped in Key Largo at John Pennekamp Coral Reef State Park hoping to find a snorkeling outfit that could take me out. My timing was good, because when I entered the small shop that advertised the service, the lady working the counter told me that a charter was heading out, and that if I wanted, I could tag along. *"Sweet,"* I thought as I signed the dotted line, then checked into the motel next door so Athena could wait the day out in air conditioned comfort.

It was clear skies for the forty five minute sail on glistening turquoise to get to the reef, which lay in about eight feet of water some fifteen miles offshore. "A perfect training ground for a guy with new sea legs," I quipped as we arrived and I could see the reef in the shallows. The charter gave a quick safety brief and told us to stay within the buoyed perimeter, then cut us loose to enjoy ourselves. Excited, I geared up fast and was the first to jump in. Our guide had reassured that there were no sharks expected, but what I failed to clarify were the jelly fish.

I had entered cannonball styles, and like a carefree kid darting off a rock at the edge of a trout filled pond. So it wasn't until I was fully submerged that I saw I was in the midst of a plume of alien's. Panicked, I fought back hard by kicking madly while also swinging my arms like fly swatters to avoid their trailing tentacles. It was a scrap I didn't pick, but still I gave a title shot performance they won't soon forget. Throughout the struggle however I was unaware that I had been splashing others who were trying to ease into the water behind me.

The Captain yelled out to get my attention, "HEY! Canadian guy! Relax! They don't sting…remember?"

"Right," I thought while giving thumbs up to show I was okay and now recalled that he had mentioned that fact earlier.

After clearing my snorkel, I put my head back under the surface to observe the ribbon of jelly's coasting by. It was a sobering sight and my defensive swat's quickly turned to endearing touch as I playfully began to push them off course and into one another. The reef was a diverse environment, standing out among the thousand shades of life were a duo of smiling parrot fish, and the odd bully of a lionfish poised with venomous spines.

Busy watching them suspiciously, I was startled when from beneath a rare armor plated relic appeared; it was a giant loggerhead turtle. She had been there the whole time I think, but because I was distracted, I didn't see her until now. As the animal moved, I followed, and from my vantage I could see every detail of her reddish brown shell. She was covered in a lean layer of algae over a pattern of scutes that resembled weathered shingles on a Victorian roof, though I knew hers were probably much older.

Lumbering whimsically with a twitching head and paddles for arms, her massive three foot wide body left a trail of soupy sediment in its wake, and left me wondering how many starry nights she had bobbed alone on the sea. When we got to the edge of the reef, she dove into the murk and out of sight, allowing me to pop my head for the first time. Seeing I had drifted a ways from the boat, and with that old fear of open water saying "get me out of here," I hightailed it back to tell the group about the encounter. It felt like another notch on the belt of an aspiring "Oceaneer."

When we got back that afternoon, the mercury was plus of a hundred degrees, and it was by far the hottest day of the trip so far. I could already feel the effects of a snorkel induced sunburn developing on my back. Knowing that a good run would perk me up and be the ideal motivation for a cold beer afterwards, I headed to a section of the state park that had interpretive trails to explore.

Stepping off from the parking area, I ran up the road until I came across a narrow muddy path that led into a dim wooded area with a dense canopy. Following it around, it brought me back to where I started and I continued further down to the next path. This one was longer and had a boardwalk that looped for about a mile over water and through the Mangrove's.

Largely abandoned, I circled the loop six times, stopping on the same bridge each pass to admire the way the foliage hung low and reflected off the stillness of a black pond. Leering into it with imaginative eyes, I hoped to see manatees, dolphins or my turtle friend frolicking inside, but settled for the occasional silver flash of a fish darting by. With my head and heart starting to pound from dehydration, and knowing I hadn't drank much water all day, I headed back to Taco to cool down.

Trying to bring my body core down, I took my shirt off, wrenched it of moisture sucked from the tropical air, then leaned forward with my palms on my knees to catch my breath. I was lightheaded, and as I watched sweat trickle off my nose to land onto the sidewalk, I grinned at the thought, *"Things could always be worse."* Closing my eyes to allow the heat to pass, I was taken back to my time in Haiti.

In my mid 20's I did a tour with the Army in Port Au Prince after the federal government there was overthrown. Haiti was then, as it is now, a poor, unstable and violent nation with few services, little opportunity, and almost no infrastructure. But it was that hot yoga heat that I remember the most because it was inescapable. Not unbearable, but stifling for sure, it was made many times worse by the weight of an Infanteers gear. Soldiers are used to pushing their bodies to the brink, it's the nature of the business, and I can recall coping with it by adopting the same slumped position I was in now after my run.

It was a land of extremes, and the counterbalance to the heat was its torrential rain that brought flooding. The city is built high up on the hills, and within minutes of a good downpour, the dirt streets would turn to rivers of mud that carried everything away. Sure it would cleanse the place of garbage and bodily waste, but it also announced a sinister side as well; unburied human and animal remains.

Victims of the political coup, the bodies would be wedged in the narrow gaps of make shift culverts, where after being exposed to the elements, they would pile in every state of bloated decomposition. Worse than dying alone or being tossed to the streets as garbage like they were, was an indignity yet to come; Port Au Princes notorious packs of wild dogs who fought for mineral rights in the form of bones.

Foot patrolling through the melee of tightly packed corrugated steel shacks called homes, we felt inclined to bury the dead we came across but realized it was a losing battle. It was a whale of a problem in some areas, and we didn't have the resources to do that work. But really, for the sake of health, safety and pride, it should have been the locals that honoured their dead, not foreign folks trying to bring security to the neighbourhood. In hindsight, I guess if they could have, they would have, the conditions just didn't allow it.

I had a lot of unique experiences there and was often taken out of my comfort zone by being attached as the third man to the Sniper Detachment. This was before I was qualified, so basically I was a glorified "radio bitch" to a couple of really experienced guys. Don't let the label of "bitch" fool you though, in the Army it's a privilege to be handpicked to work with the best, even if it means you're just carrying their extra kit. Plus it meant that while Willy, Hami, and "old condom truck" Stones were stuck working the unit's daily grind of vehicle checkpoints, observation posts and weapon's raids, I was out of sight, and so also out of mind.

The team, Scott and Karl, were typical of a trade taught to think outside the box. Well-travelled veterans of past tours abroad, they smoked cigars, wore Hawaiian shirts around camp, and did whatever they wanted. It seemed that they were immune to the regular military bull, but the privileges didn't come without sacrifice.

I can recall on one mission, we were inserted into the jungle and had to hump through the mountains under the cover of darkness to set up a hide for over-watch of our guys. A hide being just as it sounds, a prepared position, often dug in, but always covert in nature. We set up shop in a natural defilade with a rock wall to our backs and a steep slope to our front. After clearing lines of sight through the bush with a machete, we concealed our location with moss covered logs and broad leafs the size of coffee tables.

Just as we began to settle in to hide routine, it began to pour. The abode wasn't built to be waterproof at all, but its clandestine principles of were sound, and every once in a while we got proof of that. I can remember a villager stumbling onto us, and I mean quite literally, right on top of us. It was dim light, and as he stood there looking down in our direction, trying to pick up on why his intuition had stopped him, I grinned knowing there was no way he was gonna comprise the mission. Within a few minutes, the man shrugged it off and began to walk away totally unaware three heavily armed soldiers were tapping their fingers at his feet. The hide was vulnerable to the elements, but clearly invisible too.

Hide routine was always shift work. One man stayed on the gun, drawing waterproof sketches of activity, another was on radio watch and glued to the thermal optics, while the third attempted to sleep, but more often than not would usually just spend the time catching critters, like giant centipedes and tarantulas as they scurried by.

After a while the weather took its toll. Cold, wet and too bothered to eat, every bit of our skin felt waterlogged, wrinkled and miserably numb. Worse though, we couldn't even savour a smoke, because at some point they too had gotten wet. The unwritten rule in the field is that no one complains because everybody is suffering the same. So for hours instead, we sat despondent with broken cigarettes hanging from our mouths, as our spirits diminished more and more with each drop that fell.

Karl was the first to break the silence. Looking to the grey sky, then back to his front with a scowl, he whispered with the power of optimism, "I think it's clearing up fellas." Silence ensued with only our nods answering him back. Then moments later Scott followed his lead by looking to sky as well and saying, "Yup...I got a good feeling about that, buddy." He too looked back to his front to resume the silence.

A few minutes more passed, then a great salvo of lightning flashed to illuminate the valley to signal an intensified down pour. Sighs could be heard by all, but no complaints were aired. Our position began to fill with water. Staring blankly ahead, I added my two cents to the mix. "Murph." I spit the cigarette out and watched it float between my legs.

The comment forced a bout of laughter hearty enough to warm us up, until Scott, a Master Corporal, and the most senior of us, gathered himself and said forcefully to remind us where we were at. "Shut-Up...there's no talking in the hide you idiots!" We closed our mouths, but snickered periodically for hours yet, because for some reason, soldiers find it funny that things can go from bad to worse.

The memory had me laughing as I pounded back water while I stretched after the run, and soon I started to feel my body come back into a comfortable rhythm. I was hungry, but more than that I was thirsty for the beer I knew I had earned. I headed to the hotel, showered, took Athena for a quick walk, then hit a patio pub to collect. That evening I enjoyed the sunset over a feast of frog legs, catfish, gator and homemade key lime pie before retiring to my air conditioned room to crash by nine. The next day I made my way north, and satisfied that I pushed limits in Florida, I was ready to chase history again, but this time up the Eastern Seaboard.

Leg 11
Appalachia

"I have always depended on the kindness of strangers."
- Tennessee Williams

I was on an unmarked game trail near the border of Tennessee, where for days I had been crisscrossing battlefields of the Upper South in search of locations significant to the Civil War. But tired and lost after hours in Chickamauga and Chattanooga National Park, and with light fading fast on my ambition of tracing the Confederate advance against the Union Army of The Cumberland, I wondered if I wouldn't be the last casualty of that great campaign.

If I was, I reckoned, I couldn't have picked a better spot. The barrens here seemed wildly untouched, and despite being cold to my core, I stopped to appreciate them while listening for any sign of civilization that might compass me back. Hearing none though, I slapped my numb hands together for warmth, then watched my condensed breath float high into the Blue Ridge Peaks of the area's natural bliss.

The landscape had changed again. Gone were the warm flats of Florida that concealed all manner of reptilian fauna, replaced by the scent of wet leaves decaying under the patchwork of winter's mess lining the entwined roads of the frontier. Stopping periodically, I began to hear faint traffic moving in the distance, but it was obscured by the sounds of a trickling creek, singing warblers and the snap of twigs under the weight of hooved animals somewhere near.

Cutting through a copse of birch, I made my way towards its source and was relieved to find in short order a hard pack surface with signage to guide me back to Taco just as night fell. Shivering back to life inside of Taco with the heat cranked, I was sure Athena and I had dodged a bullet, or at the very least, a long cold night in the heart of Appalachia.

Settled around the 1600's, the region is rich in history and cloaked with the lore and traditions of migrants from the British Isles and German Fatherland. And while the land has always been heavily contested by some, it's prided by all for its Jack Daniel's whiskey, majestic mountains and unique backcountry banjo culture that lends well to a highland lifestyle.

The next day I drove into Atlanta and was enthused by the modern skyline jutting from a canopy of trees that hides well a metro of some 5.5 million souls. Progressive, even during the Civil Rights Movement, it has been dubbed the city "too busy to hate." Perhaps an appropriate title given that its most famous resident preached that nearly exact same thing with *"hate cannot drive out hate; only love can do that."* I am speaking of course about the late great Martin Luther King, Jr.

Born in January of 1929, Dr. King hailed from here when it was a segregated community, but rose to prominence as a voice against poverty and inequality via his nonviolent civil disobedience movement. I have always had a fascination with the man, and eager to satisfy it, I headed into the inner city to visit the Martin Luther King, Jr National Historic Site.

The complex is huge. When you enter its well-manicured grounds, you are greeted by a statue of Mahatma Gandhi, a revolutionary who influenced King with his own struggle against the oppression of the British Empire. From there I browsed displays in the visitor centre, booked a tour then strolled towards the Peace Plaza, rose gardens, and that marvelous one hundred twenty five foot colourful mural of his life.

Making my way down the street, I stopped to photograph a set of shotgun bungalows painted navy blue, complete with burgundy trim and white picket fences of Americana. Then turning to my right, I paused for traffic and crossed the road to make my way towards 502 Auburn Avenue, an average looking yellow two story Victorian Revival, with black accents, a cozy southern porch, and a small yard guarded by a thick green hedge.

Standing on the sidewalk out front I said aloud, "I have been to the mountain top…." Then, after realizing the park ranger was waiting at the open door for me, went red in the face and asked, "Sooo, this is the home where the Reverend was born?"

Nodding yes, he laughed then said, "You'd be surprised how many people do that around here."

The interior was adorned with dark stained wooden accents, wallpaper of the period and rooms that were decorated with much of the same furniture the King's used themselves. Its space looked to be frozen in time, and aware I was sharing it with a global citizen of the highest kind, I paused with eyes closed to capture the nuance. Once there, my mind was filled with the sights, sounds and smells of a chaotic young family going about their day, naive to the fact that a future Nobel Peace Laureate lived under the same roof.

Behind the ranger now, I followed him by hopping on the balls of my feet to test the floor boards, and was amused by their squeakiness as I ran my hands along vaulted walls and peered into each off-limits room. Looking to the stairs, I imagined a blossoming King coming at me full bore from the top to the bottom of the banister on route to torment his sister. That home felt like a remarkably warm place to grow up, and to me seemed like the ideal environment for a man who became everything in life, plus the ordinary.

After the tour of the house, I followed in his footsteps through the community, past Fire Hall No.6 and to the Behold Monument before venturing into the Ebenezer Baptist Church where he pastored and spread his Christian message. A recording of a sermon he gave was playing on the speakers. Feeling his passion, I sat front row and stared at the pulpit where his ornate voice once haled to calm the masses.

Once done inside, I headed out and stopped at his marble grave to pay respects and reflect on his life. Shot dead in Memphis at the age of just thirty nine by James Earl Ray, he is another example of America's tragic fraternity of assassinations. *"But what a magnificent legacy,"* I thought with a smile, romanced by his footprint and by the notion he lies in State in the same neighbourhood where he was born, grew, worked and is still adored.

After leaving town, I continued north, then east through Knoxville until I crossed the State Line to visit Charlotte and Fayetteville. Backtracking slightly, I moved south intentionally towards Cape Fear, just so I could veer north again along the coast to transit Surf City and Emerald Isle on stretches that seemed secondary to most secondaries. On the coast the weather felt warm and subtropical again, versus the chilly interior of Tennessee. So I stripped down to shorts and a t-shirt to enjoy the drive with windows down the rest of the way.

Feeling like a master of the sky after my freefall over Cape Canaveral, I headed into Kitty Hawk to visit a point that is hallowed for any wannabe skydiver like me. Once there, I found parking, then walked a path that cut through a stand of salt tolerant oak and maple trees cloaked with invasive kudzu vine thick enough to conceal the imposing feature on the other side: a single Bermuda grass-covered sand dune that rises almost a hundred feet from the otherwise flat seaside plains of picturesque North Carolina.

"Unassuming," I thought while pausing at a clearing to admire the sixty foot monument of inscribed white granite atop it. Following the path around to the right, I walked a quarter mile or so to an area where exultant figures, forged in bronze, recorded the achievements of that grandstand. Stricken by the adventure of it all, I relished that I was now in the house that Orville and Wilbur built. This is where the Wright brothers first took to flight.

It was warm and windy as I strolled circles around the effigies, speculating as I did that the conditions were much the same as they were that fateful day. Visualizing the mood, I put myself right there at the base of Kill Devil Hill with the small crowd of supporters and skeptics who had gathered and were mulling, circa 1903. Their anticipation was high, and was enough to warrant top hats and gentlemanly garb as all awaited an innovation that would soon make the world a little bit smaller.

Looking past them I could see in the distance what I know to be the clumsy forty foot wide Wright Flyer being taxied into place. Built of spruce, cloth, and cannibalized bicycle parts, it was bound together with a series of sprockets and steel bars. As a whole it presented as a cumbersome backward looking canard biplane with rudder forward and wings pushed far aft. I giggled thinking the contraption was probably highly offensive to anyone who preached the laws of physics or even common sense. But then again, there's a reason we're taught not to judge a book by its cover.

I could imagine the sound of its water cooled piston engine sputter to life as its twin propellers gained momentum then sent a plume of blue smoke into the ocean wind. Orville is assisted up, then crawls forward to lay belly down in the cockpit. Confident, but uncertain his theory will survive the test, he wiggles his body into place then gives a shaky thumbs up.

With the gesture, dubious children cheer with fists in the air as a large counter weight drops from the rear to catapult the thingamajig that's part glider, part machine, down a junction of 2x4 rails towards the drop of Kill Devil. Clearing the track the flyer launches up, then dips down, to send gasps through the crowd before sighs can be heard as it climbs again, albeit in a wobbly, undignified fashion, but nevertheless, into the realm of thin air.

Bearing towards me in slow motion, the machine lumbers by pitching side to side, and allows me a first glimpse of the pilots thick moustache, conductors hat, and slim frame being taxed by the task of steering the thing. He stays the course though, and the clunker begins to stabilize. Blocking the sun as it passes over my head, I follow it with my gaze until it begins to descend sharply to the ground, and finally comes to a skidding halt in a cloud of sandy cinders. An eternity of silence passes as the dust settles, then applause erupts as Orville rolls from his perch to cough, spit grit, and wipe his trousers before bowing to accept his hero's welcome. And a well-deserved one I'd say, right where I stood, the Wright brothers have achieved powered flight.

Okay, so at only twelve seconds long and twenty feet off the ground, the journey wasn't quite as climatic as the work the shuttle Atlantis took on, but unquestionably it was just as epic. The tale of flight is a heartwarming one full of risk, danger, and dogged determination best reserved for a high stakes movie plot yet to be shot. It capped an exceptional period of twentieth century ingenuity, and paved the way to the moon, and someday, let's hope, beyond. I walked back to Taco, changed into running gear, and added another jog to my iconic list of locations so far.

After leaving Kitty Hawk the next day, I had continued north on stretches that skirted the Intracoastal Waterways and Currituck Sound. It was full of rural communities I probably couldn't find on a map, but that I recognized. Nearly two decades earlier, my old army chum Ritchie Newman and I had passed through the same area on a hair brained voyage of discovery.

Ritchie was a tough as nails guy with a bashful smile who never complained, had muscles to spare, and was born with no fear. His confidence made for a great traveling partner. When we graduated from Battle School we had a few weeks of leave to blaze, and because I had nowhere to go, he invited me back to his family's home in Hartland New Brunswick for some well-deserved R&R.

After a night of liquid courage and boasting that included the usual drinking games, wrestling matches and triple-dawg-dares in the basement at his dad's house, we made what felt like the most rational decision of our lives and decided to go to Florida the best way we knew how, by hitch hiking there.

It was probably one of those juvenile deals where one guy introduces a game of bravado, then the other follows suit with his own to escalate, until before you know it, both are walking down an abandoned road in the middle of a winter's night sporting Hawaiian shirts, blistered feet, and a backpack full of beer...you know the kind of night I'm talking about right?

I think it's fair to say that we underestimated the distance to Florida, an easy thing to do with the power of hindsight and given that the scale on our cartoon map wasn't quite up to snuff. With doubts creeping in, but each unwilling to back down, we literally pushed forward by walking out of town and towards the U.S. border, thumbing at everything that passed our way.

Hours into the journey and things looked bleak. We hadn't a single bite yet, save for an officer in a patrol car of the Royal Canadian Mounted Police, who was working the stretch and who kept slowing down to get our hopes up before taking off again. Eventually, and probably compelled by some sense of duty, or the realization that if he didn't stop he'd have two dead kids to deal with, he pulled over for a welfare check.

Coasting up from behind us with the window down, he flashed his lights to get our attention, then asked the most pressing question I've ever heard.

"What in God's name are you guys doing in the middle of nowhere dressed for Florida?" Sober now, both Ritchie and I looked at each other with dumbstruck eyes and said simultaneously as if we had one mind, "It does seem strange..."

then peered back to the officer and began to explain by talking over one another.

"Enough! Enough," he said with palms out to diffuse the situation, before exiting the car and opening the back door like a chauffeur. Through an exasperated breath he said, "Get in fellas…I'll take you as far as the border."

"Double weird…" I thought with my best poker face, as I jumped in before he could rescind the offer. We were dropped off an hour later at the Houlton Crossing at Maine, where the nice cop waited for us to get through, then waved us off. The customs agents joked among themselves that they had never seen anyone walk across the border with a police escort before.

Somehow from there we found our way to a truck stop nearby, made a sign that read "Miami or Bust!" then sat on the curb hoping again that the kindness of a stranger would lend a hand. It wasn't long before we were approached by an eccentric middle aged man with coke bottle glasses and a tiny dog that he cradled in his hand. He had a name I could never forget, because he referred to himself in the third person as "Miles." It stumped us at first because we wondered who he was talking about.

I gather it was a label he had bestowed upon himself to give credence to all the years he spent on the road. In any case, he looked the part of an experienced driver. Wearing a sleeveless t-shirt in the cold that looked like an American flag tightly wrapped around a big belly hanging over a buckle, his blue jeans were worn and in a state of permanent slide due to his flat back side.

When he talked it was in an accent inspired by Talladega Nights, and in an inviting tone he asked, "Where y'all headin'?"

The sign I made seemed pretty clear I thought, but wondering if it was somehow limiting us, I responded, "Anywhere you're going, but nowhere close to here."

Miles revealed his lighter side with a shy snicker as he gazed to the ground, then looked up and said, "Well alright then... grab your gear... I know the way there." The impromptu introduction marked an unforgettable week on the Eastern Seaboard with escapades and a guide as good as they come.

Crammed into his cab, all three of us and the dog, slept, laughed and CB radioed our way through an ever changing Gibraltar that saw the grey of winter in the north left behind for greener pastures in the south. Miles woke us up every time we dozed off to point out his favourite vistas and haunts as he worked fourteen hour days to pick up and drop off loads in the countless big cities and small towns we passed.

In Charlotte we had to part ways. He was headed west to California, and though it was tempting to tag along, we didn't have any time left. When he dropped us off, Ritchie and I hit the liquor store, then not knowing our next move, spent the night in a downtown park, drinking beers with the locals. The next day we hitched a ride to Myrtle Beach and bummed around like bohemians for a couple of days before jumping on a thirty nine hour Greyhound ride back to the border to avoid an AWOL charge.

We never did quite make it to Florida, but Georgia and everything in between turned into one of those great periods of randomness we all have memories of from our youth. And it was all possible because of the kindness of strangers like Miles and the Mountie. I often wished I had thanked them better, but in lieu, I see that the product of their gift is that I pay it forward with strangers every chance I get.

When we got back to Canada, Ritchie and I went our separate ways, him to a posting in Petawawa, Ontario and me to Gagetown. With hectic careers to focus on, we didn't really speak much after that. I did see him a few years later though, in, of all places, Afghanistan. My unit was rotating out of theatre after our tour, and his Battle Group was rolling in to start theirs. I can remember how surprised I was to see his face.

At the time we lived in olive drab modular tents or "mods" as they're called. They each sheltered about ten guys, and were surrounded by Hesco Bastion walls of earth for blast protection. Inside they had open gravel floors, cots spaced about every three feet apart, and clotheslines that cluttered it up like a Chinese laundry. Not the Ritz by any means, the accommodation was still a welcome retreat from the sun in a place with midday heat that reached a hundred and fifteen degrees, even in the shade.

Getting ready to depart, I was packing my ruck full of the trinkets I had managed to collect, when I heard the sound of feet dragging across the gravel outside. Whoever it was, was moving towards the lads gathered around a makeshift table built of ammo crates, while they played cards from the comfort of recliners fashioned from stacked sandbags.

By this time, the camp had plenty of troops, but for some reason the sound of those shuffling feet caught my attention, and I stopped what I was doing to sit and listen. Putting my ear to the canvas wall, I thought I heard a mumbled voice say, "Lori around?" Lori being short for Lauritsen, my last name. Stoney, always the first to speak in a crowd, answered back in that bulbous voice of his, "Yeah... he's the gunner for Three Three Charlie... next mod over man. Kick the jerk awake for me would ya, he fell asleep with my chew in his pocket."

"Well that's me," I thought as I chuckled, then reached deep to pull the tin of Skoal out for a dip made satisfying because it wasn't my own. The unknown man thanked him, then crunched his way towards me again. Stopping at the unzipped door that hung closed by gravity, but that flapped occasionally in the gentle wind, the stranger revealed that he had a fresh pair of desert combat boots poking through the bottom. They didn't look tattered by miles of use like mine, instead they were new and appeared to be straight from the quartermaster's shelf. It was unusual to see such nice boots in Afghanistan and I wondered if it was a higher up out there to jack me up. My mind raced to find a reason why.

I looked around and focused on a stack of condensed milk cans that were stashed under my cot. I had been taking them from the mess hall because the Infantry guys were still being issued hard rations and my stomach was sick of them. *"But that's no crime,"* I thought. Was it because I got a jeep stuck in a ditch while chasing Hami through a part of the Camp under construction one night? Or was the stranger some rickety old Sergeant, like Holohan, who had seen Willy and I sneak into the Dutch Camp for a couple of beers when the Canadian Contingent was dry? Shrugging the last thought off as ad hoc, I burst into laughter, aware that senior NCO's don't defer such conversations when something's on their mind. I relaxed, sure I wasn't in any trouble, but wondered who then was knocking at my door?

A flick of the stranger's Zippo broke my concentration as he lit a smoke before tossing the door flap open. Against the glare of sun, his shape was nothing but a black silhouette, and although I couldn't make out his face, I could tell he was grinning from ear to ear in my direction. Saying nothing though, he stepped inside then stood there to take a long haul of his cigarette, and followed it by blowing a volley of 'O's' towards the sagging roof.

A smile came over my face at recognizing a trick I had seen a thousand times. "Ritchie...ya old dog! What the hell are you doing here?" I yelled while jumping up to force a solid embrace.

Laughing, he patted back, then typical of an army prick said, "Okay weirdo...get off me already." Pulling back, I shook his shoulders with both hands and said in return, "It's good to see ya again buddy."

We got to talking and I asked how he knew I was in country, and he responded by showing me an article from a publication called "The Maple Leaf." It was a story about Afghanistan, and on the cover was a picture of me and my fire team partner on patrol. I had no idea I had fifteen minutes of fame back home and asked if I could keep the paper. He said, "Of course, you idiot," as he threw it at me to reinforce that was the reason he had carried it all the way there.

It's strange that every time I see someone I haven't seen for a while, I'm surprised at who they are now. Listening to Ritchie tell me about the Herc flight in from Dubai, I could see he wasn't that same carefree kid I walked across the border with a few years earlier. His innocence was gone, replaced by the responsibilities of a young family man and the experiences of a veteran paratrooper fresh off a NATO tour in Yugoslavia.

It was a good catch up and before he left I stopped him with, "Ritchie! Keep your head down around here brother."

He turned to face me and said in a nonchalant way as he did with everything, "They ain't even got the parts to deal with me." We both laughed, and he pushed his way past the flap to walk away. I flew out the next day.

A couple of months later, I was working in the yard and catching up on a year's worth of chores when I heard a news flash being broadcast through the open screen door. Two Canadian soldiers had been killed and one wounded in Afghanistan. Curious, I wondered, *"From what Unit?"*

Going inside I learned that the casualties were identified as "Royals", which was short for the Royal Canadian Regiment, my Regiment and Ritchie's too. The piece showed also the sole survivor in a hospital bed being interviewed by a foreign correspondent of the CBC. Although he was heavily bandaged and didn't say a word, the way in which the soldier waved off questions about fear spoke to his character. *"Only one man could look the Reaper in the face and not care,"* I thought, analyzing his frame.

By the time I got to work the next day, the platoon was warming up for a morning run, and because everyone knew that it was brothers from the Regiment who had died, the rumour mill was in full spin. I approached a circle of buddies who were talking shop and one welcomed me by saying "Did you hear Ritchie got hit?" I nodded to acknowledge that I had suspected that, then let silence take over for a bit before speaking up, "I saw something on the news yesterday and figured it was him…you know because of the way he waves off everything like it ain't no thing." We all chuckled as I mimicked him, then we stepped off for the run, it was just another day of service in the Army.

Now passing the massive naval facilities in Norfolk as I drove towards Washington, DC, I wondered how those boys were keeping these days, then fired Willy off a quick text to let him know I was gonna be in Nova Scotia in a few weeks. He responded immediately. "You can crash here buddy." The brief exchange lifted my spirits, and I drove late into the night.

Leg 12
"I Have A Dream" In Washington D.C.

"Every new beginning comes from some other beginning's end." - Seneca

It was sunny when I arrived to the capital, but the mood felt strangely somber as I parked then hiked to the gates of Arlington National Cemetery. Once there I could see outright the FISC National Treasure that lay before my eyes. Its six hundred twenty four acres of serene rolling hills glistened intensely green to highlight the awe inspiring spectacle of some four hundred thousand white headstones. All of which were back-dropped by alluring flower beds and exotic species of plants that included yellowwood, magnolia, cedar, oak and empress trees.

Feeling enriched by the sacrifices of the warriors entombed there, I headed towards the dramatic Amphitheater, built of Imperial Danby Marble, but stopped on route to kneel at the grave of a Vietnam soldier. From the position, I read aloud his name, rank of sergeant and date of death; June 1973, then gave credence to the mystery of who he was by closing my eyes to visualize the smiling face of an All American Boy who was someone's son, brother, father, friend, and almost certainly, their shining light.

Moving up the hill I came across a marker placed by the citizens of Canada to thank American's who served in the Canadian Army during the First, Second and Korean Wars. Confused, I turned to a nearby Marine in uniform who was touring with a group of elderly. Sensing I had a question, he approached with, "Good morning, sir."

Pointing at the plaque, I stated inquisitively, "I had no idea so many Yanks served up North?"

He smiled and countered, "It goes both ways…are you a Canuck?" I raised a brow but didn't answer back, nodding only yes to allow him to continue. "Lots of Canadians have served on this side of the border as well," he said, as he pointed to the hallowed grounds, then finished with, "In fact of the three hundred sixty seven Medal of Honour Recipients buried here, nine are Canadian…bet you didn't know that?" I did not, and a bit embarrassed by it, I mumbled my thanks and let him catch up with his group again.

At the Tomb of the Unknowns, I took a seat to appreciate the sweeping view of the city as I watched for some time a keen young member of the 3rd U.S Infantry Regiment's Old Guard pace ceremoniously twenty one steps back and forth with clicking heels. Square jawed, fit, and wearing sunglasses to hide the fact he was human, his blue dress uniform, glowing black boots and decorations presented immaculately, and I could tell he put a great deal of effort into honouring those he stood vigil over.

Admiring his drill, which cast movements as sharp as violence, I snickered at children trying to rattle him and was curious if the uniforms were still just as uncomfortable as I remembered. Then while recalling the pain of an unforgiving pair of boots I once had, I shook my head and could almost hear the voice of my old boss, Sergeant Major Elms, penetrating in my head.

Calling us to attention with bellows that struck like a whip, he would inspect the ranks while issuing words of wisdom like, "Chin-up! Chest out! Suck in that gut! Look proud, look soldierly!" Then with a quick halt himself, would give the order, "By the right.... Quickkkk march!" That was the signal for the pipes and drums to lead us forward with all the pomp and pageantry of the British Empire.

It was my early twenties and at the time I was a part of a small group of Canadians who were handpicked for service as Buckingham Palace Guards in London, England. Those are the guys dressed in Red Scarlet's, tall bearskin hats or pith helmets, that are charged with protecting the Royal Family, and that stand nearly motionless for hours on end in front of the United Kingdom's most iconic institutions. I still pinch myself that I got to do that.

Living and working out of the palaces we rotated through, we had a schedule of forty eight hours on, followed by forty eight hours off. It sounds like a grind, but actually it was a perfect routine for an aspiring swashbuckler like me. During the day, we were on high alert, and so if not preparing our uniforms for the next shift, we were rehearsing emergency scenarios or doing VIP escorts. Then by night we were patrolling the properties in full battle gear, because the Brits don't take their security lightly.

Long hours, but cat naps with boots on seemed to do the trick, and besides, the rewards of forty eight hours off in Western Europe more than made up. I could venture into London, the countryside, or even to the continent on cheap flights that ran daily from Heathrow to France, Belgium and the Netherlands for a couple of pints and still be back by lights out.

The British are the best in the world at putting on a good show, no question about that, but don't let their ceremonial garb fool you, they are after all professional soldiers and on duty for a reason. Carrying loaded weapons fixed with bayonets and robust rules of engagement as back-up, they respond anytime the law is broken, or when someone is in distress. I've even seen a member chase and tackle a pic-pocket at Saint James Palace, the crook apparently naive to the fact that guards observe, then react to such cowardly acts.

Their job demands patience, so it's rare to test their resolve like that, but every once in a while, you get that one guy who crosses a line to piss them off. I can remember a time when at the Tower of London, a tourist from South Africa was acting the fool as he danced circles around me to get a chuckle. And ya, it was annoying when he started to wave his hand in front of my face, mostly because I had seen him do it to another buddy already, but hardly original, and the norm on most days.

Frustrated he couldn't provoke a response, and perhaps unaware that the public are not allowed to touch the Guard, he tugged harshly on the sleeve of my dress tunic to see if I'd move. Ordinarily the act wouldn't have bothered me much; after all, children and people trying to get close shots do all the time. But they weren't malicious about it, and this guy was. I felt liberated that he had just given me permission to speak my mind.

With a deep breath to loosen stiff muscles, I slammed my feet together, shouldered my rifle, and parried it forward with a lunge that saw our foreheads butt together. Noses touching, our eyes locked, and I commanded at the top of my lungs as the Sergeant Major had taught, "STEP BACK FROM THE QUEEN'S GUARD!"

My words felt as satisfying as the punch I wanted to throw to break his face, but the force of humiliation hit him much harder than even that. Instantly the belligerent's gaze went from brave, to grave, to white, and then to shock as he careened backwards and fell from the curb to land flat on his butt with an embarrassing thunk, flail and scream. Scurrying quickly to his feet to save face, he looked to his friends and said with red cheeks, "What was that?" I could hear the odd chuckle ripple across the crowd, as whispers of "idiot" were directed towards him from locals walking by. Nobody, not even his friends, said a word on his behalf.

Seeing the commotion, a Beefeater, that's a retired NCO, who is traditionally responsible for safe keeping the Crown Jewels, arrived to confront the man who was now making a scene. The retired NCO had seen the tourist taunting others as well, and fed-up with his lack of respect, grabbed his arm and shook him up. Once he had the idiot's full attention, he pulled him close and voiced not so discreetly, "YOU'LL NOT TOUCH OR MOCK THE QUEEN'S GUARD AGAIN...do you understand?" He finished with a quick shove to drive the point home.

Baffled that he had been called out on his own poor behavior, the man was silent, and seemed to heed the advice, because soon after his friends lead him away without any further protest. The Beefeater then turned to me, and unsure of what was in store, I braced for a blast of wrath. He began by smoothing the wrinkles from my tunic, and seeing I was tense, said, "Relax...it's half three mate." A courtesy, to let me know it was three thirty, then offered this bit of advice, "Next go round, don't be afraid to take his head off, lad."

How honoured I was to be a Palace Guard, and even to this day that ceremonial duty in the UK remains a highlight of my time with the Army. Watching the American Sentry made me long to have the experience back, but I've learned that trying to duplicate the past is never the same as living it. I rose to stand in front of the Tomb, and knowing the guard could see me in his peripheral, signaled "A-Okay" with a hand gesture, then strolled away to pay my last respects of the day.

Following the signs up a steep circular ramp that leads to an elliptical plaza, I hesitated near the top to look over my shoulder at the view of the Washington Monument, then continued the last few steps. At the top was a chained terrace that cordons an eternal flame and a bed made of grey New England slate. The modest grave markers set within were flat black and level with the grass, hardly the lofty mausoleum I would have expected. And if not for the three acres of green space surrounding the tomb, which was designed to accommodate fifty thousand people, I never would have known that this was the final resting place of John F. Kennedy Jr.

Pushing my way through the small gaggle of spectators at the chain, I squatted to get a closer look, then said aloud in his direction, "I've come a long way to be here JFK." Through the corner of my eye I could see the comment garnished strange looks from the others, but unconcerned, I continued. "I was in Dallas last month to see with my own eye's Dealey Plaza…spoke from your Podium in Houston…and I even skydived over your Space Center in Florida." Pausing, I stood to smile at those now blatantly staring then concluded our multiple crossings with, "It's been a pleasure to chase you around the country, Sir."

Washington, DC, formally the District of Columbia, is not only the centre of power for the United States, but arguably for the world. Founded in 1791 on the banks of the Potomac by once General of the Continental Army, George Washington, today it is a sprawling metro of about six million souls and home to the pedestrian friendly "National Mall," a strip of land devoted to the nation's past. Well laid out and easy to navigate, it made for an enjoyable couple of days of walking around, even as the heavens began to open up with rain.

On the second day I left Taco at East Potomac Park and strolled a short distance south past the cherry trees of the Tidal Basin and towards 1964 Independence Ave. It's an address that is a direct reference to the year's Civil Rights Act, and the spot where the Martin Luther King Memorial stands. Following the four hundred fifty foot wall enhanced by exalt phrases like, "Injustice anywhere is a threat to justice everywhere," I soon arrived to the monument and was impressed. Carved thirty feet high from granite as pure, smooth, and white as arctic ice so it couldn't fade in any light, it contained three giant slabs and was inspired by the phrase, "From a mountain of despair, a stone of hope," which was taken from his "I Have a Dream" speech.

The forward rock is a powerful relief of King, while the two aft represent the gates of struggle he passed through while alive, his strong posture being the product of that adversary. Staring off into the distance like a visionary, he stands strong in a tailored suit with arms crossed against his broad frame. He is the first African American man to be honoured in such a way on the National Mall, and as I left I felt the tribute captured him very well.

"I Have A Dream" In Washington D.C.

Moving east two miles past the Smithsonian and through the museum district, I skirted a long row of American elms sheltering tourists from the rain, and stopped at the Capitol Building for a photo, then did the same at the White House. Crossing Constitution Ave, I circled back onto the Washington Monument, then strolled the reflecting pool on route to the base of the Greek-styled Lincoln Memorial that appears as grand as any temple that belongs in the Ancient World.

Walking up its wide steps, I stopped near the top to turn back, and with Martin Luther King still fresh in my mind, stood where he stood back in 1963 and said those most iconic words, "I Have a Dream." Chills ran through my body at the power released, and I continued up the steps again. Finally to the top, I made my way through the wide pillars that support the open dim expanse of the sanctuary, and entered slowly. Initially I was blinded by daylight refracting in, but as my eyes adjusted, I could make out three distinct chambers.

The left and right housed murals of angels, principles he valued like freedom, liberty, justice, and unity, and had inscriptions of his most famous speeches, the Second Inaugural and the Gettysburg Address. In the centre chamber, under a high ceiling of bronzed girders and translucent Alabama Marble, sat, as if unmoved by time, the statue of President Lincoln.

His aura was stunning. With one hand nearly closed into a fist to show resolve during a time of war, and the other open to the tolerance that spurred it, he appears weary with disheveled hair and a bearded face that's gaunt from the stress of a long campaign. Moist eerie though was the way his eyes seemed to change expression with differing shades of angst from every angle I stood. The one constant being that they were locked in deep thought and perhaps even haunted by a regret or two. The Lincoln Memorial was to me, the most extraordinary monument I had seen so far.

After a quick bite of lunch, I headed towards my last stop in DC and the one I looked forward to most, the Vietnam Veterans Memorial. Stopping first at the bronzed statue of "The Three Servicemen" by Frederick Hart, I stood mesmerized for several minutes admiring work that spoke to me. Draped in the uniforms of the day, the young men appeared wise beyond their years, but somehow innocent as well. Their bodies are thin, taut and youthful, but clearly taxed by the oppressive jungle heat and the weight of a thousand yard stare.

It was a scene I felt I knew well. A group of infanteers on a fighting patrol emerge from a tree line and stop to confront the obstacle of open ground. Looking for a way around, they read the contours of the terrain like a book, but pause to pay homage to what stands in front of them; a mirrored black wall that shimmers even under grey skies, and that carries the names of friends who have perished. As if guided by them, I turned to approach the wall.

The structure is simple, but potent. Chevron shaped, ten feet tall, then tapering, it has inscribed in chronological order, the names of nearly sixty thousand Americans who died over there. Once on the cobble path adjacent to it, I ran my hand across the highly buffed stone in an attempt to reach each of those young souls. Then drawn by a pair of smiles, I froze to admire an intimate moment between a father and son.

The man was tall, clean cut, about my age and in a different life we could have been brothers. With his right arm, he cradled at chest level his young boy, while with his left he kept pinned against the wall a piece of white paper for the child to rub with a crayon the inscribed name beneath it. Normally people watching at a distance would satisfy my curiosity, but for some reason, and I don't know why, I felt compelled to say hi.

When I did, he turned to me, and I was surprised that he wasn't annoyed. Instead he was polite and smiled to invite a conversation. "Sorry to bother you," I said, then asked, "Do you know the soldier whose name you're rubbing from the wall?" As soon as the words left my mouth, I felt I had imposed and stuttered an apology.

Before I could finish, the man directed my query to his boy, "Whose name is that son?" Giggling, the child responded, "That's Grandpa!" His passion ignited warmth and Dad rocked him back and forth with praise for it.

"That's my father's name, he was a sergeant in the Army." The admission took me completely off guard and I wasn't sure how to respond, except to instinctively offer my condolences. Silence ensued. Sensing I was uncomfortable, he put my mind at ease with, "Don't be…I never knew him…he died before I was born."

Right then and there I felt a connection to the stranger. Both of us had known the burden of a lifetime of not knowing who our fathers were. But aware that my absent dad doesn't deserve the same breath as a war hero, I drew no parallels out loud. Instead I introduced myself and said, "I'm a Canadian tourist, and moved by the monuments of Washington." He looked in all directions, then concurred with a proud nod, and typical of every American I had met so far, welcomed me with a handshake.

Our conversation seemed easy, and he revealed that he was a tourist too, but from Utah, and on a pilgrimage of his own; this was his first time visiting his father's grave. As he spoke the closure was obvious in his voice, and he mentioned he had been to Arlington the day before. *"I wonder if we crossed paths there,"* I thought. I found myself putting his face to that of the "All American Boy's" grave I had been drawn to.

After a few minutes, and at his son's insistence, the man looked back at the wall to see what the boy was pointing at. It was then that I saw the simple beauty of the structure that was as controversial as the war itself. On the mirrored surface, and reflecting back, was the image of the father, the son and the grandfather's inscribed name written across them. The child had pointed out what we adults had failed to notice, that the wall put all three generations in same frame for the first time ever. It was touching to see, and reinforced the war still carries a legacy of pain. Not wanting to steal another moment from them, I snuck away without saying goodbye. It seemed like a sign to keep moving on.

Leg 13
"Pickett's Charge" In Pennsylvania

"A house divided against itself cannot stand."
- Abraham Lincoln

Already well into the run, I stopped at the intersection of Broad and Diamond to wait for the light to turn green. There, as with every other stop that morning, I killed the minutes by dancing like a fool to "Eye of The Tiger" as it played over and over. Noticing two black gentlemen approaching from behind, I smiled to greet them and noticed the tall one mouth the words, "Weird freaky white guy maaan..." Laughing to let him know I heard, I shrugged it off, then admitted, "I know, I know...us Canadians have got no game...." They chuckled to send me on my way.

Heading west I stopped to applaud the peaks on the "Church of the Advocate" before cutting south along a sidewalk strewn by downed garbage cans. At Girard College I circled the campus, then ran towards the quaint Market Garden district where I had stepped off from an hour before. Mustering a burst of energy as I passed Taco, I hurtled Pennsylvania Avenue, sprinted across the freeway, then bounced, with fists held high, up the seventy two stone stairs of the Museum of Art. Where once to the top, I shadow boxed in the footsteps of Rocky Balboa. Yup you guessed it, on this day I was exploring the city of brotherly love, Philadelphia.

So if after a performance like that, you're wondering, does the author of this book have a serious crush on Sylvester Stallone? Well, short answer is, yes, I do. Being a child of the 80's and 90's, it's hard not to. I know I'm not alone, and there is a reason why he resonates with the masses--the man can stir emotion like no one else. A gifted custodian of culture, he preserves its fabric by telling the tale of the blue collar folks who embroider it. And in a world of reality TV, where fifteen minute fad celebrities like the Kardashians and the Hiltons contribute nothing but stolen breath, there is value in that.

In Rocky he plays a young man on the wrong side of the tracks who gets a chance at redemption when he's offered a title shot against the charismatic and quick talking Champ, Apollo Creed. With the odds stacked, initially he lacks confidence, but takes the fight anyways realizing a lesson paramount to the American Dream; if nothing ventured, then nothing gained, but nothing lost either. Pushing Creed, and himself, to the brink of collapse, Balboa loses a close match, but goes on to cement his legacy in underdog history.

"Pickett's Charge" In Pennsylvania

The parallels in life are uncanny. Smiling as I sat to catch my breath on the top step while absorbing the cityscape, I blurted out a Freudian slip with, "Shut up already, Nick!" The outburst was directed at my older brother, and it's one that made me giggle. When we were kids, the norm for him was to wrestle me down, sit on my chest, then in front of friends, give nuggies fierce enough to leave scars. Then to add insult to injury, he would look to the heavens and in his best Stallone, yell as dramatically as possible, "Adrian! Adrian!" That of course was Rocky's famous chant as he searched for his wife Adrian in the crowd post fight. But Nick, perhaps naive to the fact that it was intended to be used by the long shot, not the bully, would torment me with it every chance he got.

In those early years we had little in common, and so blood was the only bond that obligated us. Fortunately time has closed that gap. With guidance from our mother, we learned to form our own opinions, and while separate ambitions ensured independence, there were traits that were also inherent. She herself was both composed and confrontational, and though I do enjoy the latter, I tend to lean towards the former, while Nick prefers the satisfaction of regiment and making his point well known. It seems then that we split her qualities right down the middle.

When I was younger, I was a mild achiever at best who couldn't see past the road of tomorrow, but really I had no reason to. Constantly swept away by envy of everything feign, my mind drifted relentlessly with absurd animation and fantasies. Nick by contrast was a high achiever. In school he was an honour student; for each employer, a trusted supervisor; and when it came to money matters, he saved what he earned, was the youngest client of his banker to ever buy a house, and weighed heavily every decision he made. Total opposites, for years we both admired and despised each other, and over the period I never once tried to improve on that.

I remember a time, when right after graduation, he put his neck on the line to get me a job as a labourer on the landscape construction crew he was foreman of. The company belonged to his future father-in-law, so for Nick the pressure was on, but for me, a pariah of work, it was a struggle just to be on time.

He was my ride and would call an hour or so before arriving, his stern voice was usually enough to ensure I was ready. But one morning, after a heavy night of drinking, even that effort was futile. He had rung about ten times before I rolled to answer the phone. When I did, I assured with sleep in my voice, that I was good to go. Then, and just as he had asked me not to...I fell back asleep.

He arrived a short while later, and I heard a few quick honks to let me know he was out front. Half locked in a bad dream about a headache, I rationed it had only been five minutes since we talked, so it couldn't possibly be for me. I didn't move an inch. Next he laid on the horn aggressively with blasts that grew longer with each passing millisecond. Those too I heard, but this time I snickered, visualizing some jerk outside, and dozed off again. Finally and faintly through my slumber I began to hear the strike of boot heels making their way up the sidewalk. They were followed by knocking, a ringing bell, then pounding on the door. Exasperated but still not in the game, I rolled to my side and pulled the pillow over my face.

Soon after, from the window above where I slept in the basement, I heard, "I don't believe this," followed by, "GET YOUR LAZY ARSE OUT OF BED!" My eyes opened wide. "Now that sounds like Nick..." I muttered to myself. I looked at the clock and realized it was past six, so yes, it most certainly was my brother. His ire began to haunt my thoughts, and still stuck in high school mode, I scanned a witless mind for an excuse.

"Tell him you just got sick by something you ate for breakfast…no, no…tell him a family member died… That won't work either! Say something, you moron, he's pissed!" Interrupting my thought process, he confirmed my suspicion by adding, "I'm gonna totally kill you man." Never once did Nick ever give me a reason not to take a threat like that seriously, and with muscles now tightening beyond tense, hesitantly I spun to come clean.

Just then I heard the rumble of a vehicle pulling in. Its presence distracted him so I could breathe again, and although I couldn't see who it was, I began to follow their muffled exchange. Nick was venting about me, and because he was allowed to, right away I knew it was my saving grace. Mom was home. She had just returned from a twelve hour shift at the hospital, and knowing she would be able to calm him down, I began to count my blessings.

A moment later I heard her come in, slam the front door, then rustle about in the mud room. "What is she up too?" I thought. Another moment passed with only the flush of the toilet and an odd squeak on the stairs. I couldn't tell if she was nearing, or farring, and with angst I wondered, "Should I get up to see what she's doing?"

To answer my question, the door flung open and caught me off guard. As I swung my legs to stand, I was slapped in the face by a bucket of ice water. It was a sucker punch of sorts and instinctively I threw the blankets off to rush the intruder and close the door, but slipped and fell before I could. Rolling around like a turtle in distress while trying to get to my feet, I was fed another bucket, which caused me to pause in defeat. Cold, wet and fuming at the nerve of those two, I sat in the pool on the floor and rung my shirt out before exploding, "Mom! Are you CRAZY?"

It was a typical Jude move. She wasn't a woman who awarded sympathy for laziness, instead she taught accountability the only way she knew how, by confronting me. When at sixteen I got arrested for the first time, she told the cops, "keep him." At seventeen when I broke my hand in a fight, she said, "Serves you right, you idiot." And apparently at eighteen when I entered the workforce and made light of the fact that my brother had to pick me up, she reminded me that I needed to get up. With her it was always tough love, and it was all she knew.

"Oh good, you're awake," she replied sweetly, then threw a towel at me. Tramping past, she poked her head out the window and yelled to Nick, "He'll be right there," then flicked my ear and said harshly, "Get-To-Work!" The water thing was irritating, but necessary I think, and the stinging flick was probably warranted too, but I shuddered knowing that Nick was getting the last laugh out there. With frustration in my voice I fired back, "I know, Mom…I'm up…now get out so I can get dressed!"

A short period later I slinked out to Nick's idling truck, threw my bag in the box, then jumped in the passenger seat. Locking eyes he spoke first, "Morning, cupcake…did Mom wake you again?"

I snorted his smugness away and rebutted, "I'm sure you know full well that she did." Silence eclipsed as he pulled away and not another word was said, the quarrel was forgotten by noon. Fast forward a couple of decades and the silence is gone, because Nick and I get along the way brothers should. He is a great businessman, husband, father, brother and friend, but more than that, he's still lends a supportive hand just as he did back then. I just don't take it for granted like I did, I'm lucky to have him in my life.

The weather while I was in Philly was as overcast as Stallone's movie painted it to be. But with exploring on my mind, the grey certainly didn't take anything away. From the vantage of the steps, in every direction, the skyline teemed with the activity of a worldly metropolis. And impressed by the layers of architecture and industrial character, I attempted to capture it.

Raising my phone to the horizon, I began to snap frames packed with structure. On the first I got a beautiful panorama of

City Hall, the Eakins Oval and the impressive Benjamin Franklin Parkway as it carried east towards the Delaware River. In the second I turned about to capture the pillars of The Museum, the bust of the Lion Fighter, and the iconic Converse shoe prints left by the Champ. Then I finished with a selfie at the Rocky statue.

With the Eastern Seaboard, I sensed the element of history was infused in everything I did. Philadelphia, founded in 1682, was once the capital of the nation, post the War of Independence that is. Today it doesn't have the same title, but it's still considered to be the birthplace of America. I spent a day strolling Fairmount, a gentrified neighbourhood on the North End that's dominated by colourfully painted three story row houses, brick lofts and Mom and Pop shops that butt right up against the sidewalks.

Thirsty from the run and standing on the corner of 23rd and Fairmount, I took full advantage of a small pub there called the "London Grill." Entering, the room felt alluring, it was narrow and dark, but warmed by a copper tile roof that anchored a pair of amber light shades dangling above an inviting wooden bar. Working it was a single bartender who was busy dusting rows of liquor bottles that bounced off a mirror backsplash wall.

He wore trendy black rectangular glasses, a plaid shirt, and sported a thick Viking beard to add yet another twist of culture to an already peculiar burg. On seeing my reflection he turned and bellowed as if I were a regular, "What are you feeling today?"

I hopped up on a stool, took my hat off, tossed it on the counter and answered back, "Parched man, do you have anything local to drink?"

"I do...it's our house lager...wanna try a sample?" he asked, already reaching for a glass. Nodding yes, he picked it up, flipped it, then poured a couple of ounces from the tap and passed it over. I hadn't a drop of fluid since before the run, but had salivated at the thought the whole time, so wanted to savour the moment.

Raising the glass of amber, my eyes clung to the solid white froth head as the aroma of malt caramel and toffee began to bead water in my mouth. Instead of chugging it back, slowly I tipped the glass until the brew trickled down my throat to sear with a mix of hops and citrus that quenched just as I had hoped. Lowering the glass to his smiling face, I sassed "Yup... that's the beer I want."

The next one he passed was brimful, and I thanked him for that, then asked what kind of beer it was. He answered back, "It's Willie Sutton's, of course, it's rumored he drank here you know."

"Sure he did," I rejoiced while taking a sip, then paused for a second before asking the obvious, "and who is Willy Sutton again?"

The barkeep gave a snort, and explained. "You know...the famous bank robber...Robin Hood type...prison escape artist...all that?" My blank stare must have begged for more, because he began pointing to a landmark outside. "Come on! He broke out of Eastern State, then sat where you're sitting right now for a pint…that's what they say anyways. You know the story, right?"

"Oh that Willy Sutton! Ya, of course," I replied to play along, then took another sip and added as damningly as a tourist could, "And soooo, where is Eastern State at again?"

This time he caught on to my ignorance and said, "Ahhh... not from around here I take it? It's a prison about a block away... actually a museum now, you should check it out if you have time." Taking a big gulp and signaling for another, I peered back, "I'll do that."

I resumed the trek by following his advice and made my way to the imposing feature he spoke so fondly of. Sweeping high into the sky, the building had an intimidating Neo-Gothic design with embattlements that made it look like a fortress, but in a residential neighbourhood. Cloaked by an ominous grey billow above, I approached until I was at the tower, then began to examine its thick quarry cut stone, which bulged at the seams and lead me straight to the front door.

Originally named Cherry Hill, Eastern State Penn was erected in 1829 and was the first prison of its kind to pitch principles of rehabilitation rather than punishment to its inmates. The theory was sound, take the offenders from the violence and disease of crowded congregates, give them a small segregated space to call their own, then watch patiently as they reflect on the heinous crimes they've committed with penance worthy of God's forgiveness; hence the penitentiary was born.

Called the Pennsylvania System, and praised as cutting edge, it was the model for hundreds that followed. Of course today we would call that kind of therapy "isolation." Many were incarcerated for petty crimes like stealing bread, and there a man could go years without interacting with his peers. We now know the inevitable result of such treatment was insanity.

From the office I opted to go solo on an audio tour and right away was lost in the octagonal maze of cell block spokes that jutted in every direction. The ceilings were high and almost cathedral-like to encourage prayer in a church setting. But the arrested decay of damp mortar, peeling paint, and rusted out steel contrasted the intended effect, and instead punched unease into the air. I couldn't put my finger on it, but there is something about a place where pain, fear, and despair were once common, that sends shivers right through me. Like when I toured Alcatraz, I felt like I was being watched by ghosts, and whispered, "Thanks for the beers, Willy…if you're still in the house."

As I shuffled towards the exit, I stumbled onto the mothballed remains of the cell of a man that made me growl with delight. *"Of course he's here…why wouldn't he be?"* I thought. It was clear that for some, Eastern State was an asylum, but for others it was nothing more than a pit stop on route to building a criminal empire. I had only ever been to two Federal prisons is my life, both were on this trip and at both I had crossed paths with the same notorious man. I was standing at the cell of my old friend from San Francisco, Al Capone.

The cell was as small as the others were, but had table lamps, paintings, a wooden desk, smoking chair, Oriental rug and even a cabinet radio to make the mobster feel at home. Hovering there, I remembered back to the view we shared of the Bay, then grinned wondering if I had a mug he would recognize today. Visualizing I was a guard assigned to take care of his every need, I remarked playfully, "I'm leaving for the day Al...do you need anything?"

In my mind he was sitting in red crushed velvet robe filled by a fat belly, and had hanging from his mouth a fine hand rolled Cuban Cigar. Howling buckets as he looked at pictures in a newspaper of the gang's exploits in Chicago, he paused to tap the ashes off, then leaned over to hand me his glass.

"Ya, ya Aaron...tell the next guard to bring two fingers of whiskey, the Canadian stuff I like...and crank the heat up on your way out please, it's cold in here tonight."

I beamed to oblige then said, "Sure thing, boss," and left, certain we'd meet again.

I spent the next day exploring the city's rich attractions. Then concluded by cruising the cobble walkways of the historical district. There I acquainted with the sites of an aspiring young republic, the Revolution, and the symbol of it all, Independence Hall. It was in those walls that the concept "all men are created equal" sprung to life, so I made sure to honour it right by walking the entire site.

After leaving Philly, I drove to Pittsburgh for a day, then south and east again across the Mason-Dixon on secondary roads that weaved without shoulders in and out of the wilds of West Virginia and Maryland. Inland now and well beyond the influence of the coast, I began, for the first time in a while, to settle into the quiet routine of counting bends along the open road.

Nearing Adams County, the rugged terrain gave way to manicured hills and valleys that were littered by pastures of livestock, orchards and property tree breaks that extended as far as the eye could see. Scattered among the remnants of centurial dairy farms were Victorian homes, a red barn or two, and silos that sat adjacent to wind pushed buildings left to honour pioneers from the colonial days. It was quite beautiful there and the land seemed too innocent to host a conflict of scale. But history is very clear on one thing…it did.

When the Confederate Army of Virginia, under the Rebel flag of General Lee, invaded the north for the second time, the goal was to capture Washington and end the war. Lincoln, on receiving the news, swiftly dispatched the army of the Potomac to blunt Lee halfway. The Yankee forces positioned themselves defensively at a junction where all roads met, and though the intent was to divert action with a show of force, by doing it, they actually set the stage for a titanic clash. The narrative that followed will forever be known as the Battle of Gettysburg.

I arrived at the National Military Park around noon and went straight to the Visitors Center and Museum for a bit of background before setting out. The facility was well laid out and had various displays, a timeline of events and relics recovered from the battle. Next I went to the theatre for a short film voiced by Morgan Freeman called "A New Birth of Freedom." I finished at the "Cyclorama," a 360 degree, football field length oil painting that was commissioned to recreate the epic ebb and flow of "Pickett's Charge." The "Charge" was the climatic Confederate attack on the Union lines that failed and forced Lee to withdraw from his campaign. Because of the lore it still carries today, it was the portion of the battlefield I was most eager to see.

Maybe I could locate the area with a good run. After getting changed, I left Athena in the truck with windows rolled down. Normally I would take her, but being that it was a crisp, cool afternoon, and on sacred ground, I felt more comfortable going solo, aware that her presence might offend some.

Leaving the Visitor Center, I moved west on Pleasonton Ave passed the Pennsylvania Monument, then south along Hancock where I was easily distracted by a rhythm of randomness and scenery that took my breath away. It was late afternoon, and the sun was casting orange light that pooled shadows in the contours of slopes already covered in grass. The fescue ranged in colour from electric green to shimmering gold and slammed dramatically at times with the breeze into the bounty of static statues, monuments and markers strewn about.

Moving parallel to a zigzagging split rail fence that meshed between stands of eastern red cedar, honey locust and sycamore, I crossed Plum Run and was stricken by gloom. In some realm I knew the creek below still swelled red with the bodies of soldiers from both sides who died there. Most troubling though wasn't that they endured the same fate; it was that they shared the same bloodlines too. The thought of brother slaughtering brother is a reality in Civil Wars, and it was an image that disturbed to me.

Continuing, I passed the big guns of the Peach Orchard and the Wheat Field, then came to the rocky outcrops of the Little Round Top where I paid dues to the legendary 20th Maine. Nearly out of ammunition, and knowing a defeat could pave the way to Washington, they did the unthinkable and attacked the attackers with a downhill bayonet charge.

Pacing south again along Confederate Ave, I stopped at one of the last living legends of the contest, a once mighty oak who served witness to it all. Circling the veteran's wide frame, I ran my hands over his battered bark then scanned for evidence of musket ball scars. Gnarled and sagging by the grief of time, he revealed nothing to my untrained eye and I didn't search all that long. Instead I pivoted to focus my attention on a long swath of land to my front. Burdened by its sense of melancholy, I asked the forest around me, "Was this the start of Pickett's Charge?" With a collective gust they seemed to creak, "Yes, it was."

The comprehension put a face to the cyclograph and put me in a space that until now, I had only ever read about. Intently aware that this was hallowed ground, I was careful to not disturb anything, but propped myself up against the Old Tree, then wondered what it would have been like to stand in the same spot back on July of 1863.

Concealed in the wood line by friends of the oak I stood with now, and amassing under the cover of an artillery barrage, was the activity of a mighty legion of Confederate Grey Coats preparing for battle. Where the practice of fighting was normal for these men, a sense of trepidation filled the ranks, and the jubilance that had carried their conquering army this far, was slowly beginning to fade. Lee's soldiers have a bad feeling about this day.

Pickett, the officer chosen by Lee to lead them, sees the anxiety in their eyes and draws hawkishly near to calm their nerves with inspiring words. Secretly, he too questions the orders he has received from his boss, General Lee. The advance will be nearly a mile long, uphill and through open ground against the centre of Abraham Lincoln's well dug in Union Line. It looks more daunting than promising, but bound by a code of service to get the job done, he postures like a peacock before addressing his division.

The speech is short, but to the point, and assures that their case for promoting slavery is a just cause. Stirring emotion, he uses quotes of Southern charm to stress that, to the victor awaits the spoils. Then he concludes by rattling his saber in the direction of who he calls the "northern aggressors," the comment raises a raucous roar. With nothing left to say, except to add a sigh to the fray, reluctantly he orders the formation to "FIX...BAYONETS."

The eeriness of what sounds like fifteen thousand pairs of sheering scissors ripples across the land like nails on a chalkboard. Knowing what's coming next, the Union defenders awaiting them on Cemetery Hill, Pickett's objective, stand-to and are ready to meet them toe to toe.

Pointing to the high feature with his sword, Pickett then lowers it to signal the advance as the drummers play their tunes, and General Lee salutes from the sidelines. Breaking the tree line, Pickett emerges first, and with a 2:1 advantage in manpower, he is followed by what appears to the Union defenders as an ocean of Confederate Men. Everyone involved now knows a bloodbath is about to begin.

Moving slowly to start, they are sitting ducks that walk shoulder to shoulder in tightly knit columns. And while still out of accurate range of northern Howitzers, the odd piece of shrapnel from canister shot finds its mark, and men of the Rebel ranks begin to fall. At first the carnage comes one by one, then in tandems, but then hideously, by the dozens. Colleagues, who are also close friends, rush to fill those gaps. Their spirits remain high, but the screams of the comrades they are stepping over begins to slow their stride.

The range between the two sides has closed significantly, and although Pickett's men still haven't fired a round yet, they are now subject to Union sharpshooters from above. In the mayhem I can hear the most abhorrent pops undulate and agitate my senses. Baffled, I wondered what they're from. Observing they clamor roughly when soldiers fall, I realize the sound is human chest cavities bursting like melons as musket balls from snipers pierce their inner walls. The Confederates are halfway up the slope now, and casualties are beginning to mount fast as the Fog of War thickens. Undaunted, the warriors push ahead.

Union gunners are confident they are safe from danger for now, and so take the opportunity to intensify their shelling. Using a system of bracketing to induce terror, they hurl twelve pounder shot and shell at the Rebels with an "it's them or us" mentality. Initially the cannonballs fly well over head to embed behind Pickett's men, then corrected they land ion's to the front to inflict no damage at all.

General Lee, watching from the sidelines near me, smiles that the ordinance is ineffective. Then as Union batteries check their elevation and fire again, his expression turns to shock as he realizes what is happening. The cannon balls are now landing right on Pickett's men; his boys are pinned down in a kill zone.

Chaos ensues with some of the Southerners scattering to save their own hide, while others, still allegiant to Lee and Pickett, continue to push through the melee despite the fact that they've lost half their strength. Like children flushing, then whacking gophers on the prairie, the Union, who have been on their heels the whole time, can feel the tide of battle changing and are relentless in punishing the Rebs. Pickett's Corps is in a state of disintegration and acknowledging it, he issues his famous last order, "CHARGE!"

Lead by a halo of war cries, the remaining Grey Coats sprint up the hill in a fruitless attempt to crash Union Lines. Their fighting is gallant and breaches the defences, but stalls in hand to hand combat, and weakened already by the loss of some six thousand men, they are handily repulsed. Hearing the bugle call to signal a Confederate retreat, General Lee lowers his binoculars with staunch disbelief. After years of upset victories against the much larger, much better equipped Northern Army of Abraham Lincoln, his Virginia's have finally been brought to their knees.

Watching helpless, he acknowledges in his mind that the plan was too ambitious, and that it was one battle too many. Like a guilt riddled father trying to make wrongs right, he waits for his troops to trickle into the tree line, then offers them tender, apologetic remarks. "I'm sorry, it's my fault. I'm so sorry..." he says, but deep down he already knows they have lost faith in him.

The Northern Stand at Gettysburg produced over fifty thousand casualties in just three days, but ground to a halt, at a great cost, Lee's Pennsylvania Campaign. More importantly though, it turned the tides of the war. The South was never able to recover from the loss, and when asked years later for his opinion on why the Union Forces were able to snatch victory from defeat, General Pickett responded humbly of his adversary, "I've always thought the Yankees had something to do with it."

My leg through Pennsylvania's beautiful cities and countryside gave me a unique perspective on the history that helped shape our great continent. I felt schooled by it and as I packed up and headed for the sprawling burg of New York City, I was excited at the prospect of attempting something there that would knock a long time bullet off my bucket list.

Leg 14
Live From New York, It's Saturday Night!

"Persistence pays off in dividends." - A.L.

I had been waiting in line for over an hour watching patiently as small groups of lottery winning ticket holders, and a few lucky stand-by one's too, were handpicked to pass the red ropes of 30 Rockefeller for a security screening and escort upstairs. Wanting to join the same fray, I forced a slight height advantage by rising on my toes, then teetered back and forth while waving my index finger like an abrasive flag, hoping to get the eye of the NBC Page.

Those are the fraternity of apprentices made faddish by the quirkiness of fictional luminary Kenneth Parcell from Tina Fey's 30 Rock. In the show they are portrayed as bumbling fools, but in actuality they are not, and the strict regimen of the page who approached now was proof of that. She was tall, pretty, assertive, and I thought, fit the youthful bill the Network sought.

This was my third attempt at the same vexing game, so when she came over, she did do exasperated, and with a sigh asked, "What is it now, sir?" *"I got this wrapped up,"* I thought, beaming

211

with satisfaction like I had just landed a big fish. Pulling the card of my bucket list out, I began to explain the reason I was pestering so much.

The whole time I spoke though, she had cradled across her chest a clipboard, and sported an unimpressed look that implored, "Are you kidding me?" Undeterred, I continued to deliver my message, then leaned in and finished with, "and is there any way you can get me past the line? The show is about to begin, and I did sleep outside last night."

I wasn't quite done explaining that, but found it difficult to finish over her gruff laughter, which was now turning heads. A bit red in the face as others gawked on, I shut-up to see where she was going with this. When she stopped laughing, she cleared her throat, then pointed to the long ribbon of people ahead and behind in the line, and said, "And so did everybody else my dear." I bobbed my head coolly as she stormed away, and could hear the next guy over whispering, "What a witch man." I couldn't have agreed more.

I was simply alluding to the blustery winter's eve I spent camped on the sidewalk of 49th Street. But apparently in a city that never sleeps, she had heard the line before. I couldn't complain however, the tickets were free, just in high demand because there are only around two hundred seats per episode, and only twenty one episodes per season.

To get in, you either have to enter your name into a lottery that draws from a pool of hundreds of thousands (where you're better off catching lightning in a bottle), know a member of the staff (which I certainly never will), or pursue with the same rite of passage as generations have before by doing an all-nighter like I did hoping to nab a first come, first serve stand-up from some poor chump who couldn't make it in.

Reading online that NBC issues the tic's around seven the morning of, I left the Pennsylvania Hotel, where I booked a room, at around seven the night before. Thinking that even with a stop at Times Square for a coffee and the hike there, twelve hours would leave plenty of room to spare. I didn't however anticipate that masses of teenage girls would beat me there.

Rounding the corner with swagger, and loaded for bear with a sleeping bag, coffee and Rosy the chair, who you'll remember I rescued from a desert coulee, I dropped my kit and caboodle to stop and stare. The line was already wrapped well around the building and not knowing the reason why, I asked a mother nearby. Perhaps thinking I was joking, she threw her head back, then swathed one arm around her daughter and said, "Cause my girl, like all the others, wants to see host Drake while he's here!"

"Sure she does," I mused, before asking, "Who's that?"

The lady's daughter appeared offended, then spoke in volume on the artist's behalf. "Ummmm... he's, like, a really huge rapper... do you live under a rock or something?"

I snorted to suggest that I didn't, then said straight to the fact, "More like ice. I'm a Canadian...see the chair?"

The girl didn't appreciate my humour, but the mom did and she spoke up. "Then you should know Drake is a Canuck!"

"*Checkmate,*" I thought to admit she was right, before assessing the long line with a grin and saying, "So is Murph here as well?" Nobody got the reference but me, and I didn't bother to explain the rule of irony.

With the unexpected presence of a Teeny Bop army, the odds were mounting. But as I hunkered down for a long night, I knew my sniper's patience would prevail. Soon we were well into the wee hours, and as the snow began to fall, and the temperature drop, so too did the number of willing camping on the sidewalk. It sounds shrewd, but being a person who's made best by competition, I knew that their loss was my gain, and as the frozen stragglers filed past, I embraced the merriment of shuffling into each new spot.

It didn't mean it was easy though. Like with all tests there are people who want to see you succeed, and those who will allow you to fail. At around four am, a lady, who was nice enough, and who I had talked to earlier, came over and dangled a tempting carrot. "Well, I think I'll call it a night...a couple of us are going for coffee at Mickey D's if you'd like?"

Her offer was sinless I know, but it had defeat written all over it because the truth was that I could have used a break. I was tired, my bones ached like tooth cavities, and with wet feet, I was as miserable as any. But regret not being a favourite word of mine, the thought of quitting had never crossed my mind. Recognizing the test, I declined graciously, and quivered, "Thanks, but no thanks...at a cost of only twelve hours for the rest of my life, I could shiver through ten nights."

That "buck stops here" mentality has served me well over the years, and it was only a few short hours before I too was at McDonald's. But I was doing it in style by flaunting my standby ticket and bragging that it was no work at all to get. I was minimizing of course, and in my mind I knew that my toil wasn't over yet, because now somebody had to NOT show up for me to get their spot.

Fast forward to the foyer, where I was desperate to make amends with the page who had shamed me. I was still maybe twenty spots back from the front of the line, but with mere minutes to curtains, the odds of moving up were fading. She had disappeared for a spell, but was returning now, and her brisk strut suggested something was going on. Then, and probably used to the role of villain, she broke the news.

"Sorry folks, there's no seats left…thanks for coming out." An eruption of disgruntled moans filled the air, and the mob began to move towards the door. Not one for follow the leader, I held steady, partially to avoid the rush, but also because a voice inside kept saying, "Aaron, you're not done yet."

Just then I noticed her supervisor approach with plenty of zest to stop the deserters in their tracks. House rules, written or not, were that once you leave the line you can't come back. Which was good for me, because I was now about fourth from the front. I held my breath. Scanning the few of us who remained, she leaned in and unlatched the rope, then tread in to tap the shoulders of the first few while counting aloud, "One… two… three…"

Halting at me, her strange behavior forced me to cant my head as my expression strained for logic and begged for her to come clean. Noticing barely the hint of a grin from her, I jumped high to the sky even before she could say, "you're the fourth, come with me." Not needing to be asked twice, I rushed to security, which at 30 Rock was like boarding a plane, then was steered into the elevator. Once in, the heavy doors shut without prompt and began to ascend to the ninth floor.

The sensation now felt somewhere between real and surreal, and the page, who was rude before asked kindly, "Are you excited?"

I welcomed the query and explained what I wanted to earlier. "It's literally been a long road to get here." The door opened to an empty hallway adorned by a montage of photo's ripe with satire greats.

Stepping out into it, I was stopped once more to have my I.D. checked, and to have my non-transferable, non-changeable, ticket replaced by a wristband with the date "01/18/14." Past the hall and to the entrance of Studio 8H, we were separated and handed to four different usher's. The performance hadn't started yet, but the lamps were dimming fast, and so the usher pointed to my seat, which at second row, near centre balcony, seemed incredible, then read the statute…

"Keep quiet and low when you go to your seat. You can take your jacket off, but keep it on your lap, DO NOT put it on the chair. Please don't stand up during the show, if you need to use the washroom…don't! And by all means keep your phone off, if security sees a lit device, they will confiscate it...do you understand?" I nodded yes and he slinked away.

Taking a deep breath to absorb the relevance, I watched in awe as stagehands worked feverishly to correct details bequeathed to the last moment, then hastened off as "30 seconds" rung out. It was a cozy realm that presented much larger on TV, but after years of welcoming it into my own home, it felt as comfortable as an autumn sweater to me. From my vantage I could see perfectly the main wheeled stage to my front, the music set to my left, and the prefabricated walls of original skits being whisked into place below the bright lights of an empire that Farley, Ferrell and Fallon once owned.

The place was abuzz, and as I watched the King of Comedy himself, Lorne Michaels, saunter the aisles like a general, I drifted in and out. Thoughts of a robust "motivational speaker" who's known to sport a van down by the river played out, while Fallon giggled through "Cowbell" and the always condescending voice of Seth Myers spewed a perverse version of the weekend news.

Coming to post monologue for a stirring rendition by Kate Mckinnon on the exploits of Canada's favourite son, Justin Beiber, I cheered her last jab at him. Then she lunged towards the camera and proclaimed those most iconic words, "LIVE FROM NEW YORK...IT'S SATURDAY NIGHT!"

For a half a week I had been exploring the syndicate of a city reputed to be among other things, the Capital of Crime, Graffiti, Fashion, The Yankees, Broadway, Gangsterism, and every other textile it takes to weave the "Crossroads of The World." In that time, I had crossed the Hudson to New Jersey to see the Statue of Liberty, visited the boroughs of Manhattan, Queens and Staten Island, journeyed to the observation deck of the Empire State Building, and most poignant of all, paid homage to the 9/11 Memorial.

Now after all that, and at only the cost of a single night's sleep, I was reaping the rewards of knocking a huge "SNL" bullet off my list. Laughing 'til it hurt for ninety minutes, I knew that forever and a day, I could brag about time spent in an institution of parody that has infiltrated every facet of society from late night, to the outcome of political campaigns. SNL didn't disappoint.

The following morn I left Taco in Greenwich, then wandered to Washington Square Park after a jaunt through NYU for an urban escape in one of the cities nearly two thousand green spaces. Built in the 1800's, and to celebrate the 100th anniversary of its namesakes inauguration, the park is relatively small, so has limited plantings, but is packed with structures like the Washington Arch, which I was there to see.

Made of white Tuckahoe marble, and embroidered with iconology of the president in peace and war, it's stance at the end of Fifth Avenue is imposing to say the least. Entering under it, I was greeted by a host of street performers, artists, and a man who sat on a pail busking Beethoven's Ninth from a full size piano.

From there I moved to the centre of the common area where a gaggle of tourists were posing for pictures in the massive fountain made vogue by some of the Village's most conspicuous heroines; Rachel, Phoebe, Monica and the rest of the cast of Friends. I knew it wasn't the real fountain used in the opening segment, because that one's in Los Angeles, but still I joined in by snapping a few pics.

From the fountain, and expanding out towards the tree, apartment, and wrought iron perimeter of the park, was a large circular hardscape made of paving stones. Branching off in curves from there were pedestrian paths made magnificent by their width and the presence of original benches, lanterns and statues of prominence.

It was a picturesque urban retreat, and as if all that bounty weren't enough to convince me of its pedigree, just add to the mix that it is also home to New York's oldest gallows, the "Hangman's Tree." I trekked over to the three hundred year old Behemoth to stand under its massive limbs. Once there though I knew by the will of a shiver that its name wasn't given to endear. What attracted me to the English elm was that its roots would have been privy to the Borough's highs and lows. But once there the aura was dark, so I stepped back and headed towards a shining light, an older homeless gentleman, who was sitting in a flock of pigeons.

There in the midst of the hustle and bustle of a concrete jungle, packed by the noise of eight and a half million, and where the sounds of sirens never dull, was this symbol of innocence who seemed unaffected by it all. When I approached, he was nurturing his birds with names like Lawrence and Henry, and hoping to make eye contact, I knelt down before him. His exterior was predictably rough, pale white, with a scraggly beard, and Einstein hair, he smelled like the tattered tan overalls he wore. And although his eyes showed signs of milky blindness, I could see he had a gentle, wise and unselfish side.

He smiled when he saw me hovering, and to break the ice, I pointed to a single bird, then asked, "What's his name sir?"

The man paused with his mouth open while considering my motive. Then throwing caution to the wind he lit up enthusiasm and said, "That's Steven, he's one of my oldest. Would you like to meet him?"

"Yes, of course," I responded just as enthusiastically.

"Okay, just a sec...I'll call him to you," he said as he leapt up. Once on our feet, the man grabbed both my hands and propped them with palms open, then reached into his pocket to pull out a heap of seeds. Holding his hand above mine, and like an hourglass, he slowly released his treasure until a small pyramid was left. Then leaning in close, and with shifting eyes he said, "You have to be as steady as that tree...okay?"

I rubbernecked to the Hangman and back again, then whispered, "Yeah sure...no problem."

Next he said, "Watch this," as he stepped one pace back and slowly raised his arms like Gandalf the Grey. Then without warning, he dropped them like a guillotine, releasing as he did the wrath of his feathered friends. Within seconds, the sorcery set off a frenzied melee of squawks, flapping wings, and pumping broad chests that battered my senses and tickled them raw with pricking claws all over me.

It didn't take more than a few seconds for them to clean every last seed he had spread, and just as fast as the swarm began, it was over and done, save for one pecking beast still on my head. We don't have a lot pigeons in Alberta, so this encounter was totally foreign to me. Pleased by the contact, I didn't flinch, but waited for the Bird Man to do his thing.

He reached his shaky index finger forward and used it like a perch to lure the animal off my head. Realizing it was the big one I had pointed to earlier, I began to chuckle as if I had just witnessed a magic show. The Bird Man then introduced him by saying, "This is Steven...do you remember you wanted to meet him?"

"Yes, how I could I forget," I said, impressed, while digging deep for a few bucks to tip him before I carried on exploring.

It was early afternoon as I continued west from Washington Square, then south towards Tribeca where I stopped to admire the Cast Iron Frameworks on Grand Street. From there I meandered past the shops of Little Italy, through colourful Chinatown and north into the Bowery to find traces of beatnik counterculture, punk music and ad hoc art.

The stroll reminded me of the one I did in Philly in that the layers were entwined with the persistence of gentrification, only here it was on a much larger scale. Down each tree-lined street, I saw endless five and six story rusticated stone apartments with either small steel balconies or zipping catwalks to connect the upper floors. Some had high stoops on the lower levels that were ideal for people watching, while others had quaint niche boutiques with brick exteriors, their differing tones revealed clues to the period they were constructed.

Stumbling onto a small Irish Pub on East 7th, I asked myself if I needed a beer. The taverns frontage was little more than a series of black paned windows with a green sign that read in capital white letters, "MCSORLEY'S OLD ALE HOUSE." Reasoning it was an absurd question after a long jaunt, I headed in for a quick one.

Crowded, but not packed, by a legion of patrons that were half student, half globetrotter, the interior was dingy with a sawdust covered floor, smelled of onions, chowder, chili and had memorabilia posted everywhere. Walking through and towards the counter to wait for a stool, I took up residence not far from two guys who were squabbling with the barkeeper. There were only two beers on tap, house light and dark, and the spoiled frat boys types, who for some reason took offence to that, were speaking in riddles about the condition of the place. It was annoying to hear them complain, and when the bartender had enough, he pointed to a carved mantle that read, "Be good or be gone" and yelled, "SCAT!"

I backed his assertion and found it funny because they clearly didn't expect to be kicked out. As they got up to leave, I said "Thanks for the seat boys...bundle up, it's cold out." Beside me, and howling as I spoke, was a tall clean cut lad with rosy cheeks, neatly swept hair and a distinct accent. He was serving cheese, crackers, and raw onions from a platter. Glancing at him, I grabbed a handful of the complementary appy's, then questioned, "What's their problem? I like it in here."

He agreed with body language and said, "Hey, ya can't win 'em all, right?"

"True that," I quipped as we watched the idiots walk out the door.

Quite thirsty, I asked the man serving the crackers, "Can I get a couple of those beers they were so unhappy with?"

He shrugged his shoulders but was quick to answer, "Ya sure...I guess." After putting his platter down, he motioned to the bartender by name, then grabbed the frothy mugs when they came and placed them in front of me, asking politely as he did, "Anything else?"

I ogled in all directions for a few seconds, then shrugged and said, "Nope...I'm good. Can I put these on a tab?"

He smiled with bewilderment and said, "I suppose so....ya of course you can." I thanked him and he carried on.

The beers were only half pints, and so within about ten minutes my glasses were gathering dust. I tried in vain to waive down the barkeep again, but he was too busy to do anything but ignore me. Spotting the tall server on the other side of the room, I flagged him down, and just as with the first time, he was able to secure another round. "Thanks again!" I said as he set them down.

He asked my opinion on the beers, I responded that they were good, but confessed it was the decor that impressed me more. Sensing my love of narrative, he put his cheese platter down, introduced himself as Michael, then began to extemporize. He started by isolating pieces of paraphernalia on the walls, then professed the pub, established in 1854, was among the oldest in Manhattan, and the last to allow women in.

Pointing to the foot rail of the bar he said smugly, "Do you see Harry Houdini's handcuffs? He used to frequent here." Then he attached melancholy by describing the chandelier. It was adorned by dusty wishbones that belonged to young servicemen who left for the battlefields of the Great War. Apparently before they departed, they were wined and dined here by the owner with a turkey dinner to wish them a safe journey.

The bones sat hooked as mementos while they were gone, then were snapped when they returned. The ones that remain now are from those who didn't come home. A chill ran up my spine as I conjured the scene of uniformed jubilance from 1917. Then considered the decades of despair that followed as parents and loved ones no doubt flocked here to connect to their son's last souvenirs.

"Lots of famous people have come here too, Babe Ruth, Mickey Mantle, JFK and even Abraham Lincoln to name a few...those last two were presidents, you know."

Quick to defend my intelligence, I quipped with a smirk, "Ya, ya...I know they were Mike...even in Canada everybody knows that buddy!" He put his hands up to show he meant no offence, then recommenced.

"Lincoln came in here one night for a beer, stood right on that stool mounted over there, and gave an impromptu speech about the war." We both paused, he to picture Honest Abe's apparition, and me to recall the number of places we had crossed paths so far. *"Washington, Philadelphia, Gettysburg...now in a pub with a stranger."*

With Michael still adrift, I reached for another handful of crackers off his tray, took a big swig of beer and said while crumpling the crackers into my mouth, "You guys are the nicest, most knowledgeable staff I've ever met in a bar." He agreed with a nod, and as another round came, grabbed a mug himself, raised it up, then said just as I was taking a swig, "Ya they are...but I don't work here."

"WHAT?" The admission caught me off guard, and I coughed bread crumbs across his chest.

He took another drink, brushed the crumbs off, then said as calmly as he did the first time, "Oh, I don't work here...I'm from Brooklyn. I thought you knew that?"

Fixed with a confused look I'm sure, made extra ornery by bits of food dangling from my chin, a chiliad of questions flooded my mind. Then while pointing to the platter, walls, and beer I said, "How could I have possibly known that? I just assumed...."

I waited for a response, then asked woefully, "Whose crackers was I eating? Whose tab was I using?"

He took another drink to hide a goofy grin and said, "Mine, of course."

I laughed nervously. "Why didn't you stop me?"

"Now that would have been awkward."

"*That's true,*" I thought, then offered up, "As opposed to the conversation we're having now?" We both masked our sentiments with sips of beer before breaking out into laughter. "Sorry buddy," I said while introducing myself.

He beamed back, "No worries, happens all the time. Cheers, Aaron!"

It turns out that Michael was just a friendly regular who knew everyone there, and who apparently liked to share his food with folks he didn't know, or he was just too nice to say anything. Easy to talk to, he revealed also that he was tour guide in Brooklyn and set me up on a tour of Coney Island the next day. We enjoyed a few more half pints, and I thanked for the unexpected kindness from a New York stranger, then called it a night. The tour the next morning in Brooklyn was a joy.

* * *

A couple of days later I started off at the main concourse of Grand Central Station then jumped on the Subway's Red Line to get to Harlem. On route I thought I would have to confront a naked man reading the newspaper, watch a game of "Risk" unfolding, or spot an opportune Gyro Stand somewhere, but I saw nothing of the sort.

Exiting at Central Park North, I moved along Cathedral Parkway to Amsterdam, where with growling hunger, I searched for Uncle Leo or a jolly postman to point me in the direction of a muffin top shop somewhere near. Again there was no proof of any of that. Strolling to the end of 112th, I paused in the middle of the thoroughfare, snapped a selfie with Tom's Restaurant behind me, then chuckled as I walked in muttering, "They should really call this place Monk's Cafe already."

Once in the diner I took a seat in the corner booth and eavesdropped for tidbits of Cockfights, Superheroes, the Bubble Boy, or any other number of mundane observations from the everyday. But there was none of that either. I felt gyped and a bit discouraged. Then just by chance I focused on a hanging picture of the gang clutching one another by the front door. From that I was able to finally visualize their neurotic spirits for the first time.

On the far side of the table, and opposite me, was a tall hipster dufus called Kramer, and his neighbor, Jerry. They sat quietly and listened, while on the other side an attractive woman gawked in disbelief at what George's short, stocky, bald frame, had just professed. "It's not funny, Elaine," he lashes out like a toddler after revealing his mother is now in traction because she had caught him handling his own gear.

Following more dialogue, and an explanation, George declared impulsively what no man could ever possibly commit to, "I am never doing THAT again!" he says with furious fist pumps to sell the plan. The group, far from a supportive one, brush off his promise, then begin to negotiate terms on a bet that ensures he won't last long, they seal the deal with a pinky swear.

I was giggling at the thought of the episode when the waitress approached. "Hello, sir…you look like a fan," she said.

"Was the glare of guilt that obvious?" I wondered in my head, then answered, "Ya…and I was just recalling the contest scene." We

scoffed that only in those walls could strangers discuss the issue of masturbation. I explained that I was a tourist, and that on this, my last day in the Big Apple, I was delving into the Upper West Side to acquaint with the sites of an old obsession, the 90's sitcom "Seinfeld."

I knew the explore would include a long walk to see those sites, and that I would be eating again afterwards, so I didn't stay long, and just had a plate of fries to say I was there. Continuing, I followed the symphony of honking horns south along Central Park West for a good sixty block march through the historic district. It was here, that in the shadow of thousands of jettison skyscrapers, and among the gridlock of yellow cabs, bicycle couriers and the odd horse drawn carriage, that I finally saw the New York City I had been searching for.

On my left and divided by a stone wall, lined with cobble pavers and a swath of American elm as long as the stretch I had just walked, was the sprawling expanse of Central Park. Complete with paths, bridges, ponds and ball diamonds, it was all the playground a metropolis could ever desire. I had been running there almost daily, so was somewhat familiar, but the value of its sights never dwindled, and with each visit, I felt as though I shoplifted a little more of the City's heart and soul.

On my right and in stark contrast, but not conflict, to the Park, was some of the most impressive architecture I'd seen yet. The blocks were aligned with a flush mix of high rises and cathedrals to enhance a town already known for Art Deco. And although each was built in a different era, the symmetry of all structures transitioned seamlessly and left a stunning presence of uniformity. I remember standing in awe of it all and thinking, *"I've come a long way from the desolate windy spaces of Monument Valley."* The entirety of the Central Park region made for a phenomenal course, and having experienced it intimately that day, I could understand why it is said to be the most photographed place on Earth.

I ended the walk, and my tour of New York at "The Original Soupman" on 55th, where I celebrated the show that I felt I owed a debt of gratitude to. For years, terms like, "Soup Nazi," "Festivus" and "Man Hands" dominated conversations between my step dad Len and I. The show was the catalyst that bound us when we met, and as we bonded via time spent laughing on the couch together, I realized he was a good man, good for Mom, and good for our family. I just wish they had met long before they did.

Before him, she had a revolving door of toxic relationships that included the cycle of violence, abuse, and mistakes us kids never let her forget. And it wasn't until I was an adult that I realized her struggles with men were inexplicitly tied to us. Because let's face it, even with plenty of fish in the sea, it ain't easy to attract a man when you're a single mother of three.

She hoped to find a Marlborough man like her father, a man with a ton of masculine flaws, but who was tough, stable, hardworking, and whose only mission in life was to provide. But by looking in the wrong places, all she ever got was losers who would promise riches in a big custom truck, then ask for gas money to pick her up. Some would say she was eager to please, naive, or stirred easily, but I prefer to believe that she was just testing the waters 'till Len arrived.

Clean cut, funny and reliable as hell, he was different than everybody else, and she could see that. They were colleagues at work who in short order had assumed a whirlwind romance and home together. A case study for opposites attract, they were happy until her life was cut short, but it seems even that was balanced by design, because as a guardian, when she left, she made sure we had an angel by our side.

Len's gifts are limitless. He is a teacher, not a preacher, who opened my eyes to what the measure of a man really is inside. With a humble demeanor, he is proof that men don't have to sport big muscles to move mountains. They don't have to be quick tempered to command presence, or live in a mansion to be rich. Those are facades of the insecure, not virtues I aspire to. A real man is patient, empathetic, tolerant and loyal to his circle. Len is all those, and much much more, and I am blessed to have his influence, and to call him Dad.

After I left town, I moved north along the coast of Connecticut and Rhode Island into ever older and colder communities. New York City was anything but the Gotham that Hollywood has painted it to be. On the contrary it felt clean, safe, secure, easy to navigate and made magnificent by the kindness of folks like Mike and the Bird Man. As always though, I was eager to find out what lie ahead on the open road.

An 18,000 foot freefall skydive over Cape Canaveral pushed my fear of heights to new limits.

Escaping the heat of the Florida Keys with a day of snorkeling in John Pennekamp Coral Reef State Park.

Space Shuttle Atlantis! A visit to NASA just outside of Orlando concluded my space odyssey and was the realization of a childhood dream.

Chickamauga and Chattanooga National Military Park in Georgia was just one of many battlefields I crossed paths with around the Continent.

My obsession with flight, which started in youth and was tested during the skydive in Florida, in a very real sense began here at the Wright Brothers Monument in North Carolina.

The monuments of Washington DC felt especially powerful to me. Above is the Vietnam Memorial and to the left the Abraham Lincoln.

I explored Philly with a long run in the footsteps of fictional champ Rocky Balboa, then finished at the stairs where his statue still stands.

Camping out for the night on a Manhattan sidewalk was worth it for a chance to snag a Saturday Night Live Ticket.

I spent a week lost and wandering in New York City where every day brought something fun or eccentric my way.

The pubs on the East Coast were the best on the Continent. Here at Cheers I tackle the "Norm Burger" and a pint of local ale after a day of walking around Boston.

Willy and I hitting the Lower Deck for more than a couple of pints of Keith's India Pale Ale. He reminded me that Halifax still had that hung over feeling I remembered from past visits.

Loved the charm of Old Quebec City.

Everywhere I went I enjoyed the kindness of strangers. Here I have a beer with staff after touring Steam Whistle Brewery in Toronto.

Niagara Falls at dawn in Southern Ontario.

Enjoying a well-deserved beer after crossing the finish line of my first full marathon in Grand Rapids Michigan.

The public art of Chicago captivated. Here the king of selfies himself poses in front of the mirrored "Bean" to capture the contorted skyline reflecting back.

The Budweiser Brewery is considered an institution in the Midwest.

The Gateway Arch in St. Louis was one of the most impressive monuments I visited on the trip.

As the trip wound down, and home neared, it was nice to be in the quiet open spaces of the west again.

Seen here is a typical camp site used when I wasn't staying in a big city. Note that it's just a place on the side of the road where I could pull off at days end, roll out my gear in the back, then enjoy from the tailgate a beer as the sun set on the horizon.

Home to Calgary Alas! After racking up some 30,000 miles through 38 states and 7 Canadian provinces, it felt good to finally be home.

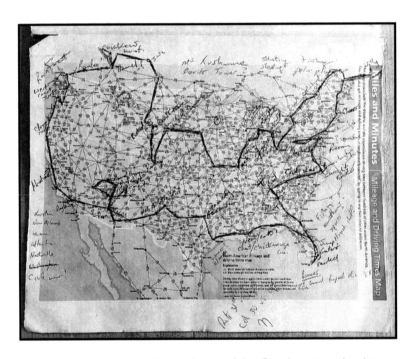

Intent on getting lost in the heart of the Continent, and using only secondaries to get there, this was the only map I used. It had no routes, but did track rough progress and was a good notepad of things to do.

Leg 15
New England Charm

"Boston is a tough and resilient town; so are its people."
- President Obama

With each chilly stroll in Suffolk County it became more apparent that moving forward, and for the rest of the trip actually, that I would be surrendering the relative warmth of the South for a land locked in an icy grip. Not that it mattered though, being a Canadian, my soul is warmed by toques and mittens on cold days, and so on this blustery eve anyways, the conditions only served to invigorate me.

I had just hiked an hour or so in a serpentine fashion through downtown and had plowed through ten inches of fresh snow to get here. Now at the top of Beacon Street, and sheltered by the canopy of the Hampshire House, I loosened damp layers to let the sweat evaporate. There as I took a break, I shook my frozen limbs awake, and stood for minute to appreciate the iconic, almost abandoned, cityscape.

Pulling my camera out to capture my lone tracks before they, like everything else, were buried by the onslaught of Toonie sized flakes swirling through the street lights, I smiled that the siege of Old Man Winter had been relentless for days, but that nothing could steal the seduction of Massachusetts Bay. (Toonies are two dollar Canadian coins, by the way.)

Off to my right was a lone cabbie stuck and spinning his wheels off Brimmer Street while he cursed bundles and threw his hands up in the air. I yelled, "Do you need a push?"

He answered back in blue collar brogue that sounded kinda Kennedyesque, "Don't botha...I got nowhere ta go anyways. Get inside and enjoy a beer for me too, please." To get here I had passed a number of perfectly good pubs on the way, but grinning that this was the one I was holding out for, I waved to oblige and followed the pointing finger on the sign saying "Cheers" down stairs.

Entering I stopped to stand beside the famous Smoke Shop Indian, who has probably guarded the landing as long as I've been alive, and surveyed the scene. The entire room was warm with oak accents and brick walls, and was dimly lit by stained glass shades with tinted decor. On those walls were caricature portraits of the cast, and badges of First Responders who frequent here, and centered above the bar was a hanging rack to make the chamber feel smaller.

Taking my toque off, I raked it back and forth against my knee to rid it of frost, then stomped my feet clean, gave the Indian a fist bump and moseyed on in. Seeing a spot at the end of the bar, I paused to look around, half expecting Carla, Woody, Sam or Cliff would be there to shout, "AARON!" as I neared, but to my dismay, no one even looked my way. I sat down on it and quipped, "Where everybody knows your name, huh?"

When the barkeep came over, I asked what Norm's usual was. He answered with a shrug, "I wouldn't know…but Sam Adams is our most popular beer." I nodded that I should have known that given his prominence around here, then said bluntly, "A pint of that, sir, and the biggest burger in the house."

He dimpled. "A Normy Burger it is."

Nursing the suds as I waited for my order, a man with an accent that painted a picture of Southie caught my attention. He was bragging of an account he had with a woman one time. Acting big and abrasive as he spoke, he sported a tight moustache, had fifty something presence and his friends called him "Peter the Mic." Based on all that, the tinge of red in his hair, his attire of slacks, a green t-shirt and a shamrock flat cap, I figured he was Irish bred.

He spoke loudly, "No, no, boys, I swear on my mother's grave…this girl was a great big burly bastard of a lass with frizzy hair and a stomach like a silverback."

I dropped my head and just about spit my beer at the unexpected candor that would be faux pas anywhere but here. I gathered myself, and gestured, "I'm good" for those who saw, then awaited the punchline as he continued.

"And I told her at the beginning of the night, I said 'Sweetheart, I got the biggest unit in town, and if you're lucky enough to stick around you'll get to see that it touches the ground." We all chuckled at the embellishment, and proved that men of all kinds share an affinity for stupid funny humor.

A moment of silence followed as he looked around, then quizzed us to get the ball rolling again. "And do you know what she had the gall to say to me after we were done screwing around?" He showed great storytelling ability, and his use of suspense kept me hanging on. "She said, 'You know, Pete, you were a bit misleading earlier…you didn't mention you'd have to lie on your belly for your junk to hit the floor."

Our laughs caused him to break character, and a few of the others approached to show they cared with a slap on his shoulder and a "There, there." Faking humiliation, he struck his palm firmly on the counter to quiet the roar, then looked around one last time and said, "I know fellas, some of us aren't so well endowed…but can you believe the nerve of that ungrateful gal?"

As if in on the joke, the bartender piped up, "And when was the last time you put your hands on her, Petey?" The Mic took a big swig of ale, burped a sigh of shame, and while sheepishly canting his head side to side said, "Well actually…that was our wedding night…so it was the last time I could even get my friggin my arms around her waist." The crowd, myself included, broke out again as he spun to make sure she wasn't in the room. He finished by explaining he loved her dearly.

I tipped my glass in his direction and he asked, "You must be married?"

I answered, "Nope...divorced."

Pointing at me with his index finger, he raised his voice, "Ahhh, but ya were! I gotta have a drink with this man before I leave here." Then he slid one stool closer to me, clinked my mug, and bellowed, "Cheers, mate." A bromance was born.

The introduction sparked a good conversation and once relaxed, one beer lead to the next, and that lead to a couple more, until we were tipsily hanging on each other's words. Apparently he was a motorcycle enthusiast and so we soon found ourselves going tit for tat with road adventures that grew fatter by the minute. He told me of past tours and places I should go, and in turn I shortlisted for him a few favs of my own. The exchange was good, and if not for the arrival of my dinner plate, we probably would have bantered all night.

Towering taller than a pint glass, the burger wobbled when the server walked, and I could see she was mindful of it. It was a handy two and a half pound double decker with onion rings, lettuce, tomato and all the fixins, plus a bundle of fries on the side just in case. I was hungry, but not that hungry, and wondering from which angle I should attack this thing, I said to the Mic, "Did I order that?"

He beamed, then nodded yes while leaning over to venture "and I don't think you got the balls to join the club by finishing it."

What he was alluding to, was the in-house Wall of Fame whose members belong to the "Norm Burger Society," an exclusive club of meat loving big eaters who have earned the title by devouring this same exact meal. Evidently I had just entered the ring and so rebuked, "Sounds like a dare."

He laughed hard and said, "Let's call it a bet…say, loser buys the next beer."

Not much of a gambler, but snickering inwardly at the prospect of racking up another stranger's tab like I did at McSorley's, I said, "You're on."

His cackle was speechless, but his body language said "well then…bring it on!" Puffing up my chest like a peacock in heat, I signaled for another drink and clamored presumptuously, "One more…old Pete is buying these." The comment garnered laughter, and as he poured it, I raised the burger, hoping that my pride wouldn't induce a cardiac episode.

The burger was good, but I'm not gonna lie, it was a rough haul and I think by about midway through I was ready to throw in the towel. The competitive juices however wouldn't allow it. As I neared the end of the road, the thump of grease induced blood pressure deafened my ears, and although I was uncomfortably bloated, and somehow covered in mustard, I licked my fingers clean to sell the satisfaction that not a morsel remained.

Pete, a man of his word, reached for the tab and said in disbelief, "Good… just add the feat to the rest of the stuff you've told me about from this wicked pisser of a trip you're on!" Then turned to me, pulled his gloves on, and said before leaving, "enjoy the Freedom Trail tommara…and welcome to Bassten."

"Boston," "Beantown" and "The Hub" are all iconic names for the historic capital of the Commonwealth of Massachusetts, and the inspiration for the famous bar from the sitcom Cheers where I had just been welcomed. Home to some 650,000 residents, many of them easy to mesh with descendants of the Emerald Isles like the "Mic," it's a burg that was literally a breath of fresh air, especially after my stint in New York, which I enjoyed immensely, but it had left me feeling a bit claustrophobic.

The drive here along the dune strewn coast reminded of the peaceful white sands I had passed through way back in New Mexico, and it was the perfect segue from hectic to corneal bliss. For the first time in what felt like a thousand miles, the metropolitan Eastern Seaboard now faded to give way to the gap of time, distance and the serenity of New England shores.

I had spent much of the drive here detoured on "The Old King's Highway" from Provincetown, where I stopped often to photograph the surrounding beaches, harbours and fishing villages sandwiched between Nantucket Sound and Cape Cod Bay. The stretch was fraught with the charm of lighthouses where fleets of whalers and sailors once ruled, and where cottages are postcard picture perfect. Refined, they had granite boulder foundations, pitched shake roofs, colourful clapboard siding and shutters to deflect Mother Nature's scorn. And if not for the patriotic zeal of American Flags flying from porches everywhere, I could've easily mistaken it for the Canadian Maritimes.

The likeness left me feeling homesick now, so I was anxious to keep moving, but determined to sightsee as well. A bit hung over the next morning, I set out to conquer the Freedom Trail with a good long walk. The trail is a two and a half mile swath through downtown that connects sixteen significant sites of the American Revolution via a red bricked, or painted, line on the sidewalk.

Parking at Boston Common, just across from Cheers where I was the night before, I stepped off and was pleased to see the snow had been cleared. Following Beacon towards Park Street, I passed the gold plated roof of the State House, until I got to the church where I paid homage to two guys I had already met, way back at Independence Hall; old Sam Adams and John Hancock are buried there. "Thanks for the beers last night, Sammy," I said while rattling a bottle of Advil in my pocket.

Continuing northbound on Washington, then east on Court, I passed the cobbles of the Boston Massacre site, then flew by Faneuil Hall. At Paul Revere's House I bellowed, "The British are coming...the British coming..." then cut through Copp's Hill Burial Grounds to arrive at the site I was looking for, Bunker Hill.

"More hallowed ground from a trip full of 'em," I thought as I stopped to take it in. It was here, and by way of Lexington and Concord just down the road, that the "shots heard round the world" were fired to ignite the War of Independence. In the

spring of 1775, the struggle was still in its infancy, so both sides jockeyed feverishly for good posturing. This feature, where I stood now, overlooked the harbour and city, so was as much a strategic objective as it was a symbolic one for all sides.

The resulting battle saw the inexperienced Yankee force strike a mortal blow against the Crown, and in doing so they proved their "Minutemen" could go toe to toe with British Regulars. Of course history records it as a defeat against the Yanks because the Militia retreated, but they did so only after they ran out of ammunition from repulsing two British charges. When the smoke cleared, a thousand Englishmen were piled high on the slopes below, and so it wasn't a bad levy for a lot the King himself called Ragtags.

His assumption wasn't unjust; the Colonials under George Washington lacked the test of credibility, and on paper didn't match up. It's no secret that the Continental Army was enervated by starvation and woefully equipped for a long campaign. But with timocracy to govern their honorable cause, and with more to lose than their foes, like the Union I observed in Gettysburg who fought so gallantly almost a century later, the men here persevered simply because their backs were against the wall.

It would have been a hell of sight to see. Kneeling at the base of the monument overlooking Charlestown, I visualized what a patriot soldier would have seen. The enemy would have been everywhere, and peering out into the harbour would have been a flotilla of His Majesty's Ships including among others, the HMS Glasgow, Lively and Somerset. The masts of those giant wooden vessels would have bobbed in the Blue Surf, but nothing could conceal the dozens of cannon they brought to support the Redcoats massing below.

Regimented, the English Infantry were the image of flawlessness and their discipline alone would have been enough to intimidate, because where one man went, the others always followed. But even more psychologically damning than uniformity was that they didn't flinch or show any fear of anything as they formed ranks of three, then marched wave after wave up the hill. It was clear as I left that the City of Boston was forged by fire.

* * *

I drove the short leg to Salem where with each mile gained, I seemed to go even further back in time. In the Deep South I was exposed to the fight for equal rights through the ravages of the Civil War circa 1863. In the Mid Atlantic it was the enlightenment of the 1700's and the War of Independence that dominated cities like nearby Boston where I had just strolled. Now trouncing down streets that were paved in the 1600's, it was the superstitious struggle that seemed to call out to me.

The area is kinda unassuming because it has a beautiful waterfront, hosts notable landmarks like the "House of Seven Gables," and affords vibrantly themed shopping in the core. But even while projecting the image of Perfectville, there were a few underlying indicators that pointed to something else.

Attractions with names like "Dungeon Museum," "Misery Islands," and "Gallows Hill" seemed to be the norm. And add to them the numerous spooky cemeteries, which were complete with giant oaks bearing scraggly limbs and raised roots as disorderly as the crooked gravestones they tossed about, and you've got a town tailor built for Halloween. Today, on my last stop in New England, I was exploring the Salem Witch Trials of 1692.

I had spent the morning at the museum opposite the steely statue of the town's founder, Roger Conant, known as "The Puritan," before venturing outside. Reeling from the displays, I crossed the street to the commons and was quickly overcome by the lore I learned there. Nobody knows for certain where exactly the executions took place, but the one constant in the area is that this park has always been here, so I figured it held a few secrets inside.

As with the rest of the community, the expanse was alluring and made inviting by antiquated features like the arched entry, paths, and a central gazebo. Surrounding it were large mansions that were built long after the fact, but that offer a good mix of nuance and layered variance. Once at the centre of the space, I pivoted in circles to contemplate, and got goose bumps trying to imagine paranoia so strong that neighbors turned on neighbors and husbands on wives.

Feeling the darkness locked beneath my feet, I conjured the ghostly scene of a horde of pilgrims chanting as an innocent young woman called Bridget Bishop pleaded for mercy. Her cries fell on deaf ears though, and instead she was dragged to the gallows, where the snap of her neck satisfied the crowd. Moving towards the eerie stature of the founder again, I winced visualizing another crime. There I could see a group of Puritans, dressed in religious garb, stacking rocks on an elderly man named Giles Corey who they were busy "pressing" to death. His crime was that he flapped his arms when talked.

Before the hysteria ran its course dozens were dead, with many more convicted of proxy offences they did not commit. And although most were eventually exonerated, I'm sure none were ever the same again. The Witch Trials felt like a grisly chapter to end the leg on, but as I fired up Taco to leave the eeriness of Salem behind, I smiled at how lucrative the Eastern Seaboard had been. Not only did I "explore, experience, then push beyond" but I left a whole lot wiser. The area's wealth of cultural and historical gems was as fascinating as any classroom I've ever been. *"Next stop...New Brunswick,"* I thought beaming that it was about time I visited my own past.

Leg 16

Coming Of Age In New Brunswick

"When we think of the past it's beautiful things we pick out. We want to believe it was all like that."
- Margaret Atwood

It was a long drive to the Border from Portland via the I-95 in Maine, but a good reflective one too. I chose to take the Interstate versus U.S. Route 1, which I had been following almost exclusively since Florida, because I wanted to revisit the crossing at Houlton where Ritchie and I had sauntered across on foot nearly two decades earlier.

Pulling up to the Guardhouse, I poked Athena awake, and as we waited, I stared up at the brilliant red and white Maple Leaf flag flying stiff in the breeze. When Athena looked at me, I asked, "What are the odds we get through this smoothly after we tell him where we've been?" As usual when I disturb her slumber, she paused, groaned, then looked the other way as if to say, "Sounds like a personal problem that doesn't involve me."

Seeing him approach, I acted like an old pro at dealing with authorities now and rolled down my window with passport in hand. The man who took it looked tired, as if his shift had started a week prior, but smiled anyways then said politely in tongue that caught me off guard, "Merci. Bonjour, monsieur... como sava?"

Old pro or not, I don't speak a lot of French in Alberta and because it had been so long since I was out east, I forgot the language of Acadia dominates many regions. Scrambling, I smiled sheepishly then replied with a shaky voice hoping not to offend, "Tres bien...ummmm...je ne sais pas parle Francais, can we speak in English please?"

He nodded yes and with a heavy accent said, to my relief, "Of course we can."

Glancing down at the passport he prodded, "Where are you going, sir, and where have you been?"

It was a question that I knew would take some explaining, so I started with a little humour to see what he'd say. "I was in California, drove along the Mexican Border, had a good stint south of Miami, and in New York too...now I'm heading to Halifax after I'm done talking to you."

He looked at his partner, then back at me with an expression that was either unimpressed or unsure of what I said, then quizzed wearily, "Have you been to this crossing before?"

I grinned. "Yup...walked across some years ago. But I had a police escort at the time." I knew it sounded trite, but in the interest of playing the game, I didn't refrain and as I expected the admission raised suspicion.

With concern written across his face, he gazed through the darkly tinted Plexiglas of Taco's canopy and whispered, "Funny guy, huh? Estee Tabernac…" He then pointed to a secondary bay and said assertively "Alright…pull in there please."

I smiled, expecting that he would have been more crass and said, "Sure thing."

Parking, I got out and unlocked everything, threw my keys on the dash then walked over to ask, "Where do the dogs go during your inspection?"

He told me to put her on a leash and just take her and grab a seat. I did, and not knowing how long the wait would be, I began to read pamphlets until after a short while the officer came over to clarify a few things.

To justify where I had been, I told him about the road trip, which absolved me of being a drug kingpin, but didn't exonerate the walking across the border thing. For that I explained, "Almost twenty years ago I walked across with my buddy Ritch, we were attempting unsuccessfully to be hitchhike Gypsies." Then told him about the Mountie who had picked us up outside of Heartland and was nice enough to drop us off here, "hence the escort part."

After answering a few more questions, that really just involved a list of locations, he started to relax and when the search was complete another officer came over with thumbs up. At the signal, he went to hand over my passport, but held back with one last cross exam, "Do you ever see that friend of yours? Ritch, I think you said?"

"We've crossed paths here and there, but lost track of one another." He bowed his head with "too bad" to give the impression he knew all too well how time does that. I grabbed my passport, thanked him, then confessed, "but the reason I'm going to 'Hali' is to see some old buds."

He smiled now and said, "Well then, enjoy the drive, safe travels and welcome home."

With temperatures hovering around -10 Fahrenheit now, it did feel like home, not like the arid Alberta I know, but more like my second home; the wet middle of New Brunswick, specifically Camp Gagetown where the Army had posted me for five years. Gagetown is not a cushy spot for the faint of heart, and because it has a reputation as austere, initially I wasn't all that excited about the twist of fate that brought me there. But like everything in life, nothing is ever as bad as we hear, and I chuckle now that it wasn't long before I grew to adore this province for the utopia it really is.

In the summer months it is as hot and sticky as any region on the continent, while in the winter it is as cold as the Arctic Circle. But for those of who have slogged through the training area a time or two, we wear the conditions with pride. We've earned the right to forever annoy with words like, "When we were young, we walked to work uphill both ways, did it in the blistering heat and pouring rain, and when we dug holes on the side of hills, they filled with swamp water to flood us out." All true, but I can feel adolescent eyes rolling already.

Although a steely place to work, it's charming too and made inviting by the resilient people of the Canadian Maritimes. They are the miners, lumberjacks and lobstermen who for eons have moved at a slower pace in this, the northern reach of the Appalachian Range. They work hard, play harder and proclaim that if Tennessee's end is the home of Jack Daniels Whiskey, then New Brunswick, Nova Scotia, Prince Edward Island and Newfoundland are the Kings of Moonshine Distilleries. The two cultures share a lot in common.

The lavish land here was for a long spell contested by the British and French in colonial conflicts like the Seven Years War, which is the reason it carries a legacy of bilingualism today. But with abundant resources to spare, anyone can see what it was they were fighting for. It's a strategic frontier that is bordered in the west by America, in the north by the shipping lanes of Gaspe, and in the east by the Northumberland Strait. Scattered along its many shores are pristine beaches and a lucrative fishing industry that harbour's quaint villages the equal of Nantucket Bay. But it's the south where you'll find one of my all-time favs, the rich blue chop of Fundy.

Characterized by towering red cliffs and thirty foot high tides, the Bay of Fundy and the surrounding Caledonian Highlands mark a noticeable change in flora and fauna that is more than just the odd whale or great white shark patrolling the surf. It is where the dense deciduous woodlands, that have commanded since I left Georgia, morph into the Boreal canopy of red spruce, balsam fir and sugar maple so symbolic of Canada.

The canopy from above is dense, but below is often open, lush and prehistoric looking with a forest floor in constant flux. Carpeted by moss, ferns, berries and lichens, every step could startle into appearance an emerald snake, salamander or bull frog. But more likely than not, it'll be the eerie sound of trumpeting moose that stop you in your tracks. It is one of the most beautiful settings I've ever seen, and I was blessed to have it as an escape, when I had time to get to the coast that is.

When I didn't though, I stayed close to home in Gagetown with quick getaways that were just as liberating. The ritual back then was to count days down until the weekend, when I could rush back to the house Friday afternoon, kick off my boots, grab my pre-packed camping gear and a case of beer, then peel away before the ex could say, "Hey, what's for dinner?"

Hitting a dirt track, I'd mud as far as my F150 could go, then conceal it under pine boughs before trekking into the woods the rest of the way. There I arrived at my fishing hole; a stream fed pond lined by rocky outcrops ideal for cannon balls and belly flops. There I'd roll out my sleeping bag, stoke a fire, crack a beer and say "salute" to the residents of a nearby family cemetery.

New Brunswick was first explored back in 1534, that's almost two and half centuries before Bunker Hill, so sites with graves of pioneers are scattered randomly everywhere, including in the middle of nowhere. Mother Nature has scrubbed the headstones clean of their identities now, but modest traces of them remain, and respecting the ground as a final resting place, I felt compelled to keep the location safely tucked away.

And I had good reason to. The action of bass, brookies and even the odd landlocked salmon were motivation enough, plus I preferred the company of none when blanketed by the heavens. Out there I could relax and get half cut from a six pack while joking with the dead. "Don't you worry, folks…even if I fished this hole for a thousand years, there'd still be ample food for you in there."

Odd behavior I know, but I was young and invincible. Why I felt so comfortable there became a mystery that grew with age, and now that I've left for good, often I reflect back, "How was it I found that spot without so much as a map?" A valid assertion given that I don't ever recall stumbling onto the pond, only that I always knew where it was.

Very recently, the puzzle began to unravel in a strange but pronounced way. I've taken to researching my family tree and learned in the process, through a biography my great grandfather wrote, that before my ancestors chased the Alberta dream in 1888, they lived for centuries as homesteaders in the middle of New Brunswick.

It was a startling revelation and an unnerving one too, because I learned also that their homesteads were only miles from where I laid roots. The discovery meant that for the years I was there, I was sharing the same soil, paths and likely even the same fishing holes they used. Provoked by the thought, I began to wonder if the blank gravestones I used to talk to belonged to my own dearly departed kin.

There is no way to know for sure, but while connecting the dots, I was flooded by memories of incidents that now feel supernatural. I recall running solo on a narrow trail I used often, and that there was this one spot I would always stop to stare, because I thought someone was in the trees. Nobody ever was.

With a chill, I flashed to an autumn eve where I drifted off to the warmth of a campfire, and was tugged awake in time to spot a beady eyed bear on the opposite side of the flames. I yelled to scare it off as Grandpa had taught, but looking around I wondered, "How did I know he was there?" And I still get goose bumps when I think about sleeping in a forest of creaking trees next to graves of pioneers, yet I felt welcome there.

A tad bit of clairvoyance perhaps, and if so it would be from my mom's side. When I was young I had an imaginary friend who for years I could carry on a conversation with, and so Mom would quip, "Aaron, I can see that you have a Guardian Angel by your side."

"Creepy, Mom," I'd say back, but I should have listened, because she had that way about her. I can cite one such occasion that made me believe there was more to her counsel than what I could see.

She was empathetic to the suffering of her patients, and had the uncanny ability to channel information about them, and sometimes even predict their deaths. One night after tossing and turning for hours from a bad dream, she came out to greet me as

I made toast to eat. With eyes bloodshot from tears, she uttered, "I think Jack died." Confused at how she could have possibly known that, unless she got a call from work, I asked why she thought that. She responded despondently, "He visited me at my bedside to say, 'Thanks Jude, I'm okay and when I looked at the clock it was shortly after three.'"

Even before I could console her, and as sure as the sun will rise tomorrow, the phone rang with a nursing friend on the other end to confirm what she had just said. "Judy, Jack died last night at around three…we're so sorry, we know you wanted to be there for him, but we just didn't have time to wake you."

Like old Jack, spirits seem to communicate in the most unusual way, but the signs are clear that they do, if you're open to seeing them. They are the shivers you get from cold air, the tunes you hum but wonder from where, the slight shadow you catch in the corner of your eye, and in my case, the guides to a clandestine fishing pond in the middle of nowhere. I'm certain it was family ties that reached out from beyond to welcome me into the places they lived, laughed and died. Who knows why, but hopefully I won't be privy to the reason for a long time.

I followed the Old Lincoln Road as it meandered along the Saint John River from Fredericton to Oromocto, the town adjacent to Camp Gagetown. I set up there for the night and was in bed early. I must have been tired, because when the alarm went off at eight, it felt as though I had just laid my head down. Still, the sun was peaking over the horizon, and not being a guy who lingers in bed, I slid out, brushed my teeth, layered well, then headed out the door for a run to reconnect with the area.

Moving east down Restigouche, then north on St Lawrence, I cut through the Q's, that's short for PMQs or Permanent Married Quarters, then rounded the turning circle, swung up Nootka Street and braked on the driveway of 25 Assiniboine. The seven hundred square foot chateau was once my home, and just like everything else I passed in the model town, it hadn't changed much at all.

Studying the property, I could almost smell the scent of my BBQ and see my good friend Hami, who doubled as my neighbor, jumping the fence for a beer. Stoney from Edmonton was the neighbor on the other side at the time, and I recall the hum of his crudely souped-up Sunfire zipping by. "The Bird of Prey" is what we called that poor thing back then. It had a spoiler the size of a picnic table, an array of decals that had no business in a small town, and a modified exhaust pipe that could infuriate the Pope. The car was an offensive attempt at sportiness in a burg that lacked trendiness. While still hung up on the car, my trance was broken as the door to my old place swung open and an occupant inside stepped out to say, "Can I help you with something?"

I smiled that I had been caught in dream land, then approached to explain. "Sorry...I used to live here. I've been out of the Armed Forces for a while, so was just reminiscing a bit." He nodded with some approval and so I continued. "You're probably not gonna believe this, but I remember a time my buddy Willy sprayed a bottle of bear scare in your kitchen on poker night and a horde of guys in regimental t-shirts poured out onto the lawn to cough up their guts…you can imagine the wife was none to impressed with us."

He smirked, looked inside to conceptualize that, then scoffed, "Some things never change I guess."

Feeling the sting of biting cold in my extremities, I picked up the pace and headed down Mackenzie towards Griffin's Pub, through the gates of the base then stopped at the theatre as my face turned red at remembering an embarrassing incident there. I and a small group of peers were called to the stage to receive an accommodation from an icon in Canada named General Rick Hillier. It was for our participation in the Cambrian Patrol and read in part, *"He represented his unit, Regiment, and the Canadian Army in an exemplary fashion,"* but in true form, it had to be presented in front of a thousand of our brethren.

The Cambrian is a long range patrolling exercise hosted annually by the British Army in Wales. It's kinda like one of those adventure races you see on TV where athletes challenge themselves to all kinds of terrain via bikes, skis and rafts. But this one is with rifles, radios, and rucksacks, and is specifically designed for the military community.

The competition is stiff, and so attracts well over a hundred teams from some of the globe's most prestigious reconnaissance, commando, mountain warfare and Special Forces outfits. All are there to vie for bragging rights, but it's a rough slog, even for that robust lot, and not long after it starts, ambitions of first place finishes turn to prayers of just finishing to save face.

I'm not gonna divulge too much, trusting that the internet can do that. But if a mission that includes heavy loads being carried over long distances, to gather intelligence, while clearing mines, rappelling cliffs, negotiating water obstacles and fighting through ambushes doesn't seem daunting enough, just add sleep deprivation, dehydration, hunger, and the guarantee of injuries to the mix, and you can be sure that the event will test even the grittiest.

On the exercise I destroyed my ankle with a snapped ligament early on. The shooting pain was incredible, almost like a toothache in fact, as I limped along on my double wide swollen leg that rolled with each step we took. But with the help of the team, I was able to push through the bogs and slopes of the Brecon Beacon Mountains to overcome. After we finished I had to have the ankle surgically rebuilt, but it was a small price to pay, because we brought back to Canada shiny Silver Medals for soldiering that was second to none. No one ever expects an accommodation just for doing their job though, ergo the red in the face reference I mentioned.

Continuing I turned right onto Champlain Street, followed it past the Infantry School and a platoon grinding out a forced march, then stopped a short distance later to pay homage to the gym where for years I spent lunch hours swimming laps inside, practicing with the football and rugby clubs at the track, and frolicking on the obstacle course out back. *"Lots of bruises gained here,"* I thought, as I rubbed my knee caps.

Crossing the street, I headed inside building "D-25" in the D-lines to warm up and again revisit an old crib of mine. The D-lines are a series of white plastered barracks with asphalt roofs that look the same on every base you go to. They are similar in principle to a university residence, but these are more like rites of passage that might just be the inspiration for movies like, "Animal House" and "Old School," because even in this strict environment, they are refuges for chaos. The troops affectionately call them "The Shacks."

Looking around I snickered knowing that if the walls could talk, they would tell tales of guys wrapped in towels, and skateboarding down flights of stairs on triple dawg dares. Of innocent gatherings turned into week long benders with "Shack Rats," a kind of soldier groupie, and of how friends find the most bizarre ways to invade one another's space.

I can remember a time, when after an all-nighter downtown, that I walked up to my room around six am with key in hand and ready to crash, but found someone had beaten me to the punch. The door was open just a smidge, but enough to allow cigarette smoke and Saturday morning cartoons to billow out. I knew it wasn't my roommate Moe, because he was deployed to Africa, so I was baffled at who it might be.

The room had nowhere to hide, it was barely twenty by twenty feet wide, had a dotted hospital style floor, fluorescent lights, and a radiator that had to be drained twice to work. All in all it was quite a utilitarian abode. For us it didn't matter, we didn't own much anyways and had only two single beds, a tubed TV, and a beer fridge we called the "Admiral" because she looked like a 1951 Airstream Winnebago.

Her generator-like thrum and ability to chill beers to about one degree less of frozen, then bead them with sweat when cracked open, was a perfect companion to bachelorhood. In front of the Admiral and facing away from the door was our brown leather sofa that had been slashed, battered, and tattered without due process. But that we had gotten free, so never really cared who squatted on it, until now that is.

I pushed the door open slowly and could see the top of the man's head bobbing to the antics of Bugs Bunny and Friends. With dark hair and dark skin, it clearly wasn't Moe, and conferring my concern aloud, I said to let him know I was there, "Who is in my room, Bro?" The culprit didn't move an inch to acknowledge me.

Circling, I stepped gingerly forward and laughed realizing the man was no threat, but that it was just my buddy Dave from down the hall. He was wearing my housecoat, slippers, and a pair of my dress socks, but nothing else. Still glazed from a night of saucing himself, he somehow had the strength to sip, through a straw, a bottle of vodka he found tucked in my drawer.

"Good morning, Davey-Boy," I said to interject.

He looked up with googly eyes squinted, and beamed as though I were late. Then with a slurry Cape Breton accent that coupled humourously with hiccups, said, "Yes indeed, by'e…it is a great mornin'!" His instinct was to get up and greet me, but by the way he was sitting spread eagle I could see nothing but junk was violating my robe, and so as he peeled his bare bottom towards me, I pushed him down.

"You stay right there, big guy, you're in no condition to say hello."

He furled his mug, shrugged, and burped to agree, then silence engulfed the room as he resumed his breaky of fermented wheat. Wanting to get to bed I muttered, "Where are your clothes at?" He raised his brow, but didn't answer. I pressed the conversation again, "And how did you get in, buddy? Thought I locked er' up?"

His response to that question was quick and seemed straightforward enough, "Oh, you did…I had to come through that crawl space over there," he said, pointing to the eighteen inch square window that I now noticed lacked a screen to guard against New Brunswick's mosquitos.

Walking over to inspect, I poked my head out and peered past the six inch nano ledge that runs around the entire second floor, and as I did observed, "No ladder here." Then I saw twenty five feet down, my screen, which was peeled and tossed like a used tin of sardines. Pivoting to address it, I said "Well, there's my screen all crumpled on the ground."

He motioned with a wink and a grin, then replied through those persistent hiccups, "Yup, and you're gonna have a hard time explaining that to the NCO's."

I nodded and said, "That's true, they won't find it as funny as you do," then broached the obvious again. "Where are your clothes at, Dave?"

He smirked this time as if I were a fool, then canted his head to give a clue, took another sip and said "They're in my room."

"Yes, I gathered," I fired back, deducing that he would now justify the act by saying he shuffled along the ledge to my window for a good reason. Once there, he somehow managed to get his big frame in, found it funny to get changed into my clothes, then moseyed on back to his room through the door to drop off his attire, before returning to sit in wait for the ultimate story. That would have been the rational version I think.

Instead he looked at me straight, and with a stone cold face said, "I was out on the ledge, right…couldn't get back in, and so shuffled here to see if you were around."

Picturing what a traffic stopping scene it would have been to watch a hairy man do such a thing, I asked one final nagging question, "And you were naked the whole time?"

"Yes," he said, as if he were under oath.

Not bothering to clarify why he was on the ledge naked in the first place, I admitted, "Ya, that makes sense," then sat next to him, grabbed the bottle, and took a pull of the vodka.

I winced and gave it back. As I did, he raised the elephant in the room, "And now I'm locked out…can I crash here till the maintenance guy comes around?"

I shrugged "Ya, of course…sleep in Moe's Bed for all I care, he'll never be the wiser."

Dave responded, "I already did."

Warmed up by the memory of what passed for normal back then, I headed out to finish the run. Moving towards my old stomping grounds, and where India Company's building used to stand, I stopped to let a column of LAV's roll on by. Those are the workhorses of the mechanized infantry. They are sixteen tonne armored vehicles that move like greased lightning on eight wheels and bring to bear a gauntlet of weapons systems that include twenty five millimeter cannon.

It's a machine I knew well from the all the courses, exercises and deployments I spent filling every role of the section, from riflemen, to driver, to gunner and crew commander in. As the last vehicle passed, I gave the sentry thumbs up and mouthed, "Good vehicles." The crew, as if they could somehow hear me over the roar of their Detroit diesel engines, honked back to concur.

On my way back to the hotel, I cut through the Sergeants Mess and halted one more time to recall Hami's last minute wedding inside. It was less than seventy two hours before we deployed to Haiti, and despite our protests to the contrary, Tom, who had a newborn child at home, was fearful that if he was killed his girlfriend and child wouldn't get the benefits they deserved.

It wasn't our first tour, so initially we brushed it off as nonsense to convince him otherwise. But Hami was adamant, and so Willy and I dressed to the nines in scarlet uniforms to stand up for him. And as always, we celebrated in grand style with shots of Drambuie as we danced like idiots late into the night, then paid the piper with hangovers that lasted for a week after we got to the jungle. Although it seemed a little inconvenient at the time, I'm glad I have memories like those today. Just add them to the book of so many I have from this place.

Service to me is the highest calling, but I know it doesn't appeal to all. It seems the path is reserved almost exclusively for youth from the lower socio economic classes who lack the opportunities that rich kids get. So in that way, I was blessed by being from a broken family, because it felt like my destiny.

If you ever need to verify the impact the experience had on me, just take a look in my front closet and you'll see enshrined the only proof I have left, that once I rolled with the best of the best. There hangs the tunic that I pull out at least once a day to remind myself of it. Befitting of a former Buckingham Palace Guard, its brass still shines even in the dark, and its cloth is pressed crisp and consummate to bear the weight of medals awarded for tours abroad. As a whole it appears to be in limbo, and in a high state of readiness, should one day I get the chance to dawn the uniform of my nation again. Hey, a guy can dream, right?

I know I make light of a lot in life, but you probably understand that not everything in the military was the gravy train of beer drinking and finding naked guys. The organization does the heavy lifting of the country, and it's a dirty business where lives are put on the line, so the professionalism of its members is undeniable. It's just that when I reflect back, I can only remember the good times, because they are the ones that still make me laugh till I cry, call them the salt of army adversary.

The jaunt around the camp offered a reconnect to a place I know I would be lost without. When I arrived here I was a punk kid with a terrible attitude and few tools to whittle it down. But circled by high achievers and positive mentors, I learned a rule I now live my life by, "You are what you surround yourself with." This place was truly my first coming of age.

Leg 17

Nova Scotia's Ocean Playground

"Remember that the most valuable antiques are dear old friends." - H. Jackson Brown, Jr.

It was day two in Halifax, and relieved that the bitter cold of New Brunswick was behind us, for now at least, my old buddy Willy and I were out on the town to exploit the mild weather passing through. Already we had banged out a ten mile run through Point Pleasant Park, toured the Keith's Brewery downtown because it's an institution around these parts, and crossed the harbour from Dartmouth a couple of times in search of Celtic bands.

As an outsider, the industrial cityscape of bridges, hills, and landmarks, like the Clock Tower and Citadel Hill, appeared Bostonian in nature. Partly because of their close proximity on the coast I guess, so influences are inherent, but also because of the similar Colonial style buildings, mixed with modern skyscrapers and swaths of colourful wooden homes inspired by Queen Victoria herself.

Like Beantown, "Hali's" quaint shopping districts are walkable, within range of public gardens called "Commons," and are usually only a stone's throw away from a waterfront boardwalk littered by tall ships, and a world class pub scene frequented by patron saints who answer to MacDonald, McVicar or McDade.

And something else I noticed, although it's not on everyone's list, is that both burgs sport spooky graveyards chalk full of prominence. In Halifax, for example, you can go to the old Fairview Cemetery for a picnic with the ghosts of the Titanic if you wish, just keep the jokes about the cold water to yourself. I'm not sure they're ready to make light of icebergs just yet.

The bond that runs on the surface is also one that was solidified by catastrophe. In December of 1917 tragedy struck here when a French cargo ship, called the SS Montblanc, which was carrying munitions bound for the war in France, collided with a Norwegian vessel in the narrows of Halifax Harbour.

The resulting explosion released 2.9 kilotons of energy, and was the largest man made blast ever released prior to the Nuclear Age. I can't imagine the sound of that, or what the force of waves rippling across the water would have felt like. But I suspect a baseball bat to the chest would be an understatement, because the city was completely leveled by it. Each time I walked through, I felt the residual of that dark day, and although I tried to visualize it, I couldn't seem to replace the beauty of now with images of chaos then. An apocalyptic scene for sure, it would have been fraught with survivors searching through ruins, rubble, and heaps of snapped trees for loved ones.

Within hours of receiving word of the disaster, the good people of Boston organized, gathered donations, and loaded a train with people, material, and medicine bound to help with the grisly task of recovering eleven thousand dead and wounded residents. It was a selfless act that we've never forgot, and one that's still taught in our history classes. To show appreciation, each year the provincial government of Nova Scotia sends on behalf of its grateful citizens a giant Christmas tree to stand lit in Boston Commons for the holiday. It's the reason why I think the two cities felt like sisters to me.

It had been ten years since I was last here, but it seems I was picking up right where I left off. With a dozen beers pounded back after a night of catching up, I was just about to get kicked out of the pub for drinking too much. Well, Willy was anyways, and true to form he sealed the deal with words of wisdom from, "Talladega Nights." Giggling like a schoolgirl as we were being escorted out, and I'm paraphrasing here, he stops then slurs something to the effect of, "No, no...wait, wait, wait! I'm a diabetic and a veteran...can't we stay for just one more drink?"

The act did little to impress the doorman or change his mind, and in a Garrison City like this, I'm sure he heard the plea a million times. But being the nice bouncer that he was, he explained patiently before pushing us towards the door, "Sorry lads...you haven't done anything wrong, except cross the sober line, so ya gotta keep moving on."

"Fair enough," we thought as we looked at each other and shrugged it off. There was, after all, no disputing the fact.

By 3:00 am we were set up in lawn chairs in Willy's driveway and smashing back the assurance of a hangover the next day, when I leaned in and asked, "Why do you always pull the diabetic' card?"

He snickered, then garbled with a smidge of shame in his voice, "I don't know man...I just like being an idiot when I get all Willied-up." That much I already knew, and conceded that some things never change.

A former army guy who today is a highly decorated member of the Royal Canadian Navy with multiple deployments abroad, he is an anomaly. Known to be a perfectionist at work, an attentive family man at home, and a marathoner everywhere he goes, he is hardly a diabetic, but the term "all Willied-up" does apparently still carry clout.

Getting us kicked out doesn't mean he's a bad drunk though, because he's not. On the contrary, with an inviting smile and contagious laugh, he's harmless. It's just that when I was young and looking to unwind, Old Willy was my partner in crime, and he never once disappointed back then, or for that matter, now. Perhaps the lecturer Ralph Waldo Emerson said it best with, *"It is one of the blessings of old friends that you can afford to be stupid with them."* I couldn't agree more, and it was good to be in the company of an old friend who felt the same.

With a lull in the conversation, we began to doze off under the cloudy sky, then Willy shook me awake to say, "Hey...don't forget...tomorrow we visit Tommy-Boy in Wolfville for the day!"

"How could I forget?" I thought, nodding back to acknowledge him. The meeting had weighed heavily on my mind the whole road trip, and for some reason I felt a sense of dread by it, though I didn't know why, and so I didn't mention it.

There was no reason to be anxious. "Tom Hamilton," "Tommy-Boy," "Hami" or just plain old "Hams," was for years one of my closest pals, and the third man in our clan of Three Stooges. Easy to be around, he was both soft spoken and outspoken, and we shared countless experiences from work, to play, to being neighbors back in Gagetown. So I knew I was being irrational. Taking a deep breath to calm my nerves, I fired back, "I'm looking forward to it...I just don't know how I will react."

Willy smiled the words away and said, "Ya, well, just relax...it's only Hami." I agreed that I knew as much, but still, when we went inside, I had trouble falling asleep.

When I did awake, it was later than expected, but the booze had run its course. Willy, who I could hear up playing with his infant son for some time, poked his head in and said, "Hey, you getting up or what?"

Not gonna lie, I was a hurtin' unit, and it was an effort just to roll over to answer him. Once I did, I struggled through blurry eyes to see where he was, then spoke with cotton in my mouth, "Yup...Halifax has still got that same hung over feeling I remember." He laughed, and I dragged my sorry self out of bed for a shower. Within an hour we were out the door.

Coming for the afternoon was his wife Emmy, their son Wesley, and because Tom is a native son of Nova Scotia, a case of Keith's India Pale Ale. The drive was only about ninety minutes, and we hoped it would cure our pounding headaches. It didn't of course, but the conversation on route more than distracted from the pain, and it wasn't long before we were reminiscing about the glory days.

We talked about practical jokes that always went too far, and I told him about the trip, and how the American Southwest reminded me of that time we moved all of Hams kit outside in Afghanistan. Proudly we recalled leaders who left impressions from our army days, men who were carved from stone like our old Commanding Officer, then Colonel Jonathan Vance. He used to have us dig our own graves when we were "killed" on exercise, then make us stand in them while the battle raged on. Tough, but fair, he was a soldier's soldier who was loved by his troops. At the mention of his name, we both paused to smirk at the character he built.

The memories were flowing as freely as the draft had the night before, and the more we talked, the more I wanted to know. At one point I asked, "By the way, which of you two clowns put the dead fish in my truck that time?" The question triggered silence and returned only a glare. I was impressed that even after all these years he kept the secret to himself.

I was referring to a prank where I was the sole victim after Hami helped me pepper Willy's Chevy Malibu, or "Boo" as the boys called it because it had Old Man appeal, with fluorescent paintballs. So the fish was more revenge than anything, but as always it came at the worst possible time, and in the most sinister way. While away on an exercise, I made the mistake of leaving my truck unlocked. Not totally unheard of in a place with no crime, but the naivety did have me question my faith in mankind. Because unbeknownst to me, Willy, along with Tom, who was obviously playing both sides, hatched, then executed, a very cunning plan.

It was a three week stint, and so when I returned, I was exhausted, filthy and looking to get home. Walking across the Parade Square towards my truck with my rucksack, I paused halfway noticing that something was wrong. The windows of the Old Girl were all fogged up, instinctively I muttered, "What have they done?"

Approaching with caution, I opened the door, then stumbled as a waft of putrid air slapped me off guard. It smelled like fish, but with death attached, and had a musky scent many times worse. I realized then that while I was away struggling to make ends meet, someone had placed a rotting carcass inside my ride. The pickup was parked in the hot New Brunswick sun, so the gases being released were the source of the fog. *"Gross,"* I thought, trying to summon enough courage to man up.

After sitting on the asphalt for a minute, I got up to do just that. The first place I checked was the glove compartment. There

I found a nice note that read, "Welcome home, sweetheart." Written in block letters it was clearly not from my wife. Next I checked the console and pulled another note that said, "Getting warmer, cupcake." This one had a smiley face drawn in to make me shake my head. Then I reached my hand under the seat to feel around, and that's when I almost lost my lunch.

From the heat, the remains of the fish had broken down into nothing more than the slime of bio material, oil, and mushy pulp, whose aroma had penetrated every inch of the carpet. Swallowing hard, I pulled it out and stared at it for a moment as it dripped throughout. Not mad though, I was actually stricken by the genius of it and began to giggle because I knew the target wasn't me at all, it was the person I'd have to explain it to. *"The wife is gonna be pissed,"* I thought, giving them credit.

And understandably she was, because for one, that smell was never gonna come out, and two, I should have at least warned her before she got in to drive to work. I'll never forget the look of hate in her eyes when she came back inside and said through gritted teeth, "Tom or Willy?"

I made light, because I knew that I was partially to blame, then answered back, "Both, I think."

The trip down memory lane was exactly what I needed to calm my nerves, and riding shotgun for the first time since I left Calgary, I was able to truly get lost in the scenery we were passing. It was a region we had traversed together before, but many years earlier while taking part on a bicycle tour. During, we hit all of the province's gems from the north, to the south, and back again. On Cape Breton Island, where homes are painted to look 'Tartan,' we ground out the steep grades of the Cabot Trail, visited Fortress Louisbourg and then Sydney Harbour. Afterwards we headed all the way down to where we were now, and ventured into the Annapolis Valley for a solid leg.

Then as now, the entire region seduced me with its rich culture and geological bliss. Referred to as "Canada's Ocean Playground," it is a place where you will never be more than forty miles from a rugged shore line of cliffs, boulders, salt marshes, sandy beaches and idyllic fishing villages with names like Lunenburg, Peggy's Cove and Pleasant Bay. Lost in the beauty of it all, the drive to Tom's took no time at all.

When we pulled up to the church in Kings County, the butterflies in my belly were working overtime. But as soon as I saw that beautiful granite headstone in the shape of a cross fit for an Angel and inscribed "HAMI" to prove one resides there, I felt instantly calm. The humble abode is Tom's final resting place, and I could sense his energy was there to invite me in for a beer, but awed by the surroundings, I took a moment to breathe the air.

It was a spot he had, in part, handpicked for himself in the event he might fall. And if you stop just shy of where he lies now, down the road about a half a mile, it's easy to see why. There you will find a viewpoint that overlooks the valley's vineyards, orchards and dairy farms that have the quaintness of the Pennsylvania countryside.

It was the ides of January now, and so the landscape was windswept and winter kept under a layer of snow. But the smell of salt still rung true, and if I listened carefully to the breeze whistling through, I could almost hear the sounds of the bagpipes that have haunted this place for four hundred years. Unlike New Brunswick, which is heavily forested, here the land had been cleared forever. So off in the distance I could see that most beautiful feature, the mud flats of an inlet attached to my old stomping grounds, the Bay of Fundy. "Nice place to hang your hat, Tommy-boy," I said, smiling with envy as I approached.

Tom was killed in the Arghandab District of Southern Afghanistan in December of 2008. He was the most senior of three Canadian men killed that day, and it took place just three days before he was scheduled to return home from his third combat tour in that Theatre. Superstitious soldiers know better than most that the rule of 'Murph' plays no favours in the desert.

He died during a period of intense activity, and in a region known to produce high NATO casualties. Though you wouldn't know it from news reports at the time, because back home the media was consumed with reporting the hardships of celebrities like Britney Spears. Thanks to their diligent efforts, the masses prayed for the safe return of the hair she shaved from her own head.

Meanwhile, spread thinly across the lunar landscape of the Arghandab, and living on the fringe with sand storms, camel spiders and blistering sun in tiny forward operating bases, or on Provincial Reconstruction Teams, were the weary members of the Royal Canadian Regiment Battle Group grinding out one grueling day at a time. They were infanteers like Tom who volunteered to fight there, but I can't help but wonder if our North American culture has forgotten what heroes really are.

Tom was one of the many unsung heroes cut down in the prime of his life over there. With an athletic build, movie star looks and an all-Canadian smile that could make girls blush, he was a complex character to say the least. The softer side of him was the loving father to a little girl, and the one who manned the barbecue while others mingled with a brew, and the guy who wore his Beret crooked each day, got jacked up for it, then laughed as the sergeant walked away.

The flip side though, and what the outside world saw, was a professional soldier with a piercing eye that was born of a sum of close calls. Hami was a warrior of the first order who could have stood toe to toe with any Roman gladiator, and it's fair to say that I admired him very much, and wept when Willy called to break the news.

Standing there at the foot of his grave as Willy swept the snow away to reveal a shrine of un-cracked beers left by army chums on previous visits. I smiled, realizing why I had dreaded this moment for so long. Hams was like a little brother to me, and the guilt of not being with him when he died had been eating me up inside.

Because I had left the service just before that last deployment, I knew there was no way I could have been there. But what tormented me were scenarios of the way he died. I hoped he had gone quickly in a blast, but didn't know for sure and so I visualized the worst. I wondered if he was scared, if somebody had held his hand to see him off, or if a friend who was present had written home on his behalf. The images were much more graphic than I ever care to admit, but know that his passing was hard for me to accept.

Plus after all the experiences we shared, I couldn't make it to his funeral. Calgary at the time was locked down by a blizzard that grounded air traffic. And so being five thousand kilometers from where he was buried, attending wasn't in the cards. But Willy, as he always has, stepped up to the plate. The two of them were always the closest in life, so it seems fitting now that he saw him off to the other side.

It was a military funeral of course, but Willy helped with the planning, acted like a liaison to the family, was on the bearer party, and made the speech to kick off the party. He even pinned Tom's medals on his chest before they lowered him into the ground. I can't tell you how hard that would have been to do.

Soldiers who die abroad don't die nice deaths surrounded by friends and family in cozy hospital beds. They die in the dirt they lived, where one minute they are as strong as an ox, but the next, reduced to a heap of burnt flesh, shattered bones and severed limbs. Then when they come home, they don't come off the plane looking ready for a wake. They come off in a steel box, packed with ice that melts on the long trip back. Can you imagine what it's like to pick up your best friend, a person you have shared a million beers with, and all you can feel is his body sloshing around as you carry his coffin across the tarmac?

For Willy's strength, I tipped my hat, grabbed a cold one, then raised it to the both of them and said, "Cheers, boys!" The act felt like reconciliation, and without a word more being said, we pounded them back. As I did I could almost feel Hami patting me on the back with, "I'm doing good buds."

The stress of the reunite was all in my head and it closed another chapter on the trip. We headed back to Hali soon after for a feast of scallops, mussels, biscuits and beer at the "Economy Shoe Shop" on Argyle Street. The next morning was day eighty of the voyage, and as I left town to head west, it felt like the trip was winding down. But I knew I still had lots yet to cross off my Bucket List before I could say that.

Leg 18
La Belle Province, Quebec

"Bite the bullet, and it will surely open doors moving forward." - A.L.

Sitting in the bay window of a coffee shop as I enjoyed a hot cup of Joe, I reflected, while waiting for my order of poutine, on the beauty I was surrounded by. For the last couple of days I had been driving west, and following the tidal waters of the Saint Lawrence River inland towards the heart of the continent. Along the way we had encountered one blinding blizzard after the next in what was quickly turning into a winter wonderland. But the temperatures weren't cold, and so I took advantage with snowball fights targeting Athena-Bear at rest points. I think she enjoyed the game as much as I did.

The scenery along the stretch was breathtaking, and even with vast lengths of wilderness back-dropped by cliffs to look at, often I found my gaze drifting out into the Seaway in search of the U-

Boats that once stalked it. In World War Two that same water was under siege by the Nazi's, so I knew sunken ships of the conflict were not far out of reach. The tidbit added captivation to a region fraught with history, but I knew I was just scratching the surface, and that the real charm lie in the eastern townships I was transiting through.

They were dominated by cozy homes and picturesque "Mom and Pop" boutiques that greeted me with fine wine, cheese, croissants, and a whole spectrum of other cuisine.

I felt out of place, and as if I were in Europe, not North America, but still I tried my best to blend in. It wasn't easy though, especially given that to express words, I had to talk with my hands, which left me kicking myself for not taking French way back in school.

Feeling a sense of national pride envelop, I smiled at the man who dropped off my order and said, "Merci, Monsieur." As he walked away, he laughed and I turned to focus on the blue "Fleur-de-lis" flag blowing in the wind, then mumbled before digging in, "Ahhh, La Belle Province, how I'm loving thee." Today I was exploring the greatness of Old Quebec City.

First settled in 1535, the place had Old World rapture that was locked in a harsh winter. Its architecture reminded me of New Orleans a bit, but was clearly in a league of its own. It was a fortress actually, and the only one of its kind left north of Mexico, I was impressed at how well it was preserved.

Most buildings inside the ramparts are constructed of rock masonry that will never crack, and that were pushed right up against the sidewalks. The roads were pedestrian friendly, and defined by handsome granite curbs and cobble pavers that have clucked beneath horse-drawn carriages for hundreds of years. And everywhere I looked I saw cathedral's, bell towers, copper roofs, statues, and castles like the "Chateau Frontenac." Everything had ledges that bulged with snow, and the city resembled Santa's North Pole.

Walking along the twenty foot high walls that run its perimeter for miles, I ran my hands over antiquity and was stricken by what was hidden within. I knew the stones were witnesses to the continent's turbulent past, and that the cannons at The Citadel spoke to that. *"This was the site of The Battle of The Plains of Abraham,"* I thought, acknowledging the significance of a conflict fought even before the War of Independence.

The battle marked the conclusion of the Seven Years War, and settled the question of whether we would speak English or French moving forward, I'd say the Brits got the better of that debate. But the wonderful legacy left by what was then the capital of "New France," is French culture still seen in the eight and half million residents who live in the province. Everything about Quebec felt foreign to me, and made for the perfect platform to "bird-watch."

As I've aged, I've become more of an introvert than an extrovert, and so my contentedness comes from quiet outings meant for two, or just one. And if I'm alone at a coffee shop or pub, like I was now, I enjoy bird-watching to kill the time. My birds though being the unsuspecting human types that fly by, always in a hurry, and always unaware that they are being admired from afar. I'm certain that by now, most readers of this book have figured out I'm a bit of a people watcher.

In fact I've mastered the skill so well that I can safely conceal my gaze while searching for hints of a stranger's character, and as I do, they're none the wiser. It doesn't take long for a narrative to unfold, and often forgetting where I am, I'll find myself smiling as they smile, or frowning as they frown at the sight of a text coming in from who I suspect is a lover or a friend.

But that kind of daydreaming can come at a cost too like it did that afternoon as I wolfed down my poutine. I was following a man as he crossed the street, but I gave myself away by laughing when he slipped on the ice. The funny part wasn't the fall, because that looked painful, it was the awkward dance he did trying keep his balance before he succumbed to gravity and landed flat on his butt.

As I howled, all eyes in the cafe pivoted onto me, and I realized I had come off as a complete jerk. "It's not funny," I quipped defensively to those who would listen, "it's that I did the same thing yesterday." The admission took some of the pressure off, and the gawkers let me be. But turning back to view the man struggling to get to his feet, I smiled knowing the statement was only half true, because my slip was actually a whole lot more nerve racking than his, and actually it almost ended my trip.

I had been staying at a hotel in Levi, that's the community on the opposite bank of the river. Once up and around, I headed out the door and towards the ferry to cross to where I am now. It's

Quebec, so snowflakes the size of Montreal had been falling for days, and a fresh foot of white stuff blanketed everything.

Following my GPS directions, I cut through a residential area, then followed the "Cote du Passage" around until it arrived at a steep hill. The street tapered to two lanes, was undivided, and as it descended towards the ferry, it curved to the right. There was something about it that triggered my spidey senses, and concerned, I pulled over to make an assessment.

The hill was slick and looked treacherous because the snow hadn't been plowed yet, but others were able to negotiate it, albeit in a fish tailing fashion, so after engaging Taco's four wheel drive, I figured I'd be fine. I waited until they got to the bottom, then put the stick in first gear, popped the clutch, and slowly began to roll ahead. The grade was deceptively steep and dropped faster than I had anticipated. Even Athena raised her head to watch and stare as we crept past the point of no return.

Although I could feel the wheels sliding a smidge, the first thirty yards were a cakewalk, and I reckoned soon Taco's tires would find a rut and dig in. Another of my naive assumptions on this trip I guess, because with the thought, I hit the ruts, popped in to them, then back out again, and began to slide sideways where I occupied both my lane and that of oncoming traffic.

There were three vehicles behind me, and I could see they were faring about the same, and because we maintained a constant speed, there was no danger in my mind. I tugged at the steering wheel to guide the truck, but the act only seemed to make things worse, and about halfway down, I had to admit aloud, "I don't think we can stop now."

I was the lead vehicle in a runaway train that was quickly gaining speed, which was not good, but things were about to go from bad to worse. At the bottom of the hill, a driver decided to do a last minute U-turn, and so the three vehicles behind him were now boxed in too. Effectively the man turning had just set up a roadblock for a seven car pile-up.

Instinctively I began to honk my horn to warn them, but because they were boxed in, they couldn't get out of the way. Desperate to avoid a collision that might cause injuries, I cranked the wheel again to try and send Taco into anything, even a street light would do the trick. This time the tires did exactly as I asked them to, plus much more. They bit into the snow alright, but instead of taking me off the road, I was flipped another 90 degrees, and was now careening backwards down the slope. Face to face with the guy behind me, I muttered the only words I could: "Screw you, Murphy!."

I was totally helpless, and could see in my side view mirror that a collision with the stopped cars was imminent. Bracing for it with white knuckles, I was jolted violently just before we hit, and realized the jarring sensation was from the curb. Apparently as the road curved, I continued on my straight path, crossed both lanes and the sidewalk, then slammed into a snowbank. Relieved I had stopped, but with heightened senses, I followed the other cars as they piled into one another. My heart was pounding, and I remember thinking, *"That was another close car wreck."* The heap of smashed up cars reminded me of an accident I had as a young man.

At the age of sixteen, I bought my first car from a guy named Jim in an exchange of services. He was retired, well off, and so didn't want cash. But because he owned a large acreage west of Calgary that needed fixing up, he put me to work. The deal was I get the car, and he gets a year's worth of chores from me doing things like digging up weeds, cleaning his shop, and painting black that thousand foot asphalt driveway of his.

It amounted to a lot of elbow grease, but it was also a no brainer and worth every blister, because a 1979 Pontiac Parisienne, in mint condition, was a hell of a prize back then. The car was a huge "Two Toned Bullet" that was brown and beige, and built as solid as a tank. It even came complete with an eight track, shiny spoked rims and a V8 engine that consumed gas like it was going out of style. Plus because of the heavy suspension, it floated like a boat and turned heads. I fell in love with the idea of owning that car, and because Jim had let me take it home for motivation, I sat nestled in its crushed velvet seats each chance I got, listened to the radio, and dreamt of cruising for chicks.

As soon as I cleared the debt, and with the ride officially mine, I was eager to set out on a maiden voyage. More than anything though, I was proud of what I had accomplished. By earning the car independently, it had more value to me than the free hand-me-downs my peers were getting from their parents. And I wasn't the only one who noticed the hard work. My car was the only one in the household, so after the drama of getting kicked out of school a couple of times, now I was able to redeem myself. It meant Mom could drive to work instead of taking the two trains, and two buses she was accustomed to. And because she was going to drive it too, she helped get the insurance set up, which was nothing more than liability, but made the Bullet street legit.

Fast forward just two weeks after that glorious day. Me and the boys Moochy, Leaf and Scotty-Dawg, were out celebrating the new found freedom. We had been drinking beers in a park, not much though, I only had two, and them maybe a couple more. But like idiots afterwards, we piled into the Bullet to cruise about.

Stalking residential areas, we looked for something to do, and stunted with bursts of drag racing on every straight away we came to. The Beast was speed resistant, but when the petal was to the metal, all of us cheered the sound of the engine roaring, and loved the feeling of the front end lifting off the road like a speedboat cutting through water.

A little too confident after a few rips like that, I stopped at the top of a hill to rev'er up good, and the boys encouraged me to drop the foot, none of us were wearing seat belts of course. The layout of the ground was almost exactly the same as it was this day in Levi. There was a steep slope that curved to the right before leveling out, except then it was summertime, late at night, and in a residential neighbourhood.

The speed limit of the area couldn't have been more than thirty miles an hour, but by halfway down I was easily going double that. Entering the curve, the Behemoths wheels began to slip, squeal and screech, causing a jolt of fear to engulf me. I could see I was quickly losing control, and to correct, I cranked the wheel as hard as it would go. As I did, the whole car wobbled and vibrated from the stress of G-forces, then began to lean to one side as that massive suspension worked overtime to keep us from rolling. Collectively we let out a gasp of "Whoa, Whoa, Whoa!" and cringed at what we saw was coming straight at us.

Veering off the road, we hit the concrete curb, and because we were going so fast, it acted like a ramp that launched us. Sailing through the air, we cleared a four foot hedge, and I could feel Scotty and Leaf in the back grab a hold of my headrest as one of them yelled, "Brace yourselves, boys! Brace yourselves!" In turn I let out the first thought that popped in my head, "We're dead..." It was one of those "crikey" moments we all have when our lives flash before our eyes, and we think we're gonna die.

After going airborne, and flying for a distance that would make the Dukes of Hazzard envious, my beloved "Bullet" smashed right through the brick wall of a garage and into some poor guys Camaro inside. The impact, and sudden stop was harder than anything I've ever felt, and because no one had a belt on, we were tossed like dominos inside. Scotty got the worst for some reason, he flew from the back seat and went head first into the windshield to leave a soccer ball size punch in it with his skull. The sound of the collision was as loud as a bomb blast, and its energy rippled across the sleepy neighbourhood, within seconds porch lights were springing to life.

Frozen with fear, and pinching ourselves to see if we were still alive, I began to contemplate the sight. From my perspective, I could tell we were partially buried under a pile of rubble because the roof had collapsed and the crumbling bricks, mortar and drywall from the building were pouring into the shattered side windows like sand through cracks.

The car was embedded midway up the wall at the same elevation and angle we soared in on, and because debris was pinning the accelerator down, the wheels spun at full throttle as the motor smoked, wined and was ready to blow. In the air I could smell gasoline and burnt electrical, whether it was from the house or car I don't know, but I could also hear the sound of sirens near as spectators in slippers and housecoats circled the car.

Although we had lots of bumps and bruises, by some miracle not one of us was dead. We were severely dazed however, and it took some effort to shimmy ourselves out. When we did, and once the adrenaline passed, Scotty collapsed. With authorities inbound, the instinct was to run, and so I did, at first anyways. After bolting from the scene, I made it maybe a block before my Angel called me back with, "Aaron, you gotta face the music on this one, or the course of your life's gonna change in a bad way."

Knowing I had to bite the bullet on crashing my Two Toned Bullet, regardless of what the penalty would be, I turned around, walked through the crowd, checked on Scotty who was still down, then approached an officer and said, "I ran because I was scared... but this is my fault, and it's my car."

He squinted at the admission, nodded, then said, "Come with me." I was taken to his cruiser, then back to the District Three Police Station for questioning.

Once there, he asked, "What were you thinking? You could have killed a dog or something."

I was still shook up, and locked in introspect as the events played out over and over in my head. With a quivering chin, I lifted my head and answered him, "or even a kid." This veteran cop, who no doubt had witnessed so much in his career, and who dealt with so many punks like me before, must have seen I was being sincere, because it felt like from that point on, he took me under his wing.

After I passed the breathalyzer, he gave me a good talking to and reinforced how fortunate I was. Then said, "I'm not charging you with anything Aaron because you're not legally intoxicated, but you and I both know you shouldn't have been driving, and you should know how much grief you caused tonight."

He told me that the family whose house I hit, had just moved in a week earlier, and that they had saved for years to buy a property of their own. The fact hit home, because after saving for my own now written off treasure of a car, I knew what it was like to want something so badly it hurt. Then he said, "And your Mom's upset, I just spoke to her, she mentioned she'll be taking the bus for a while again?" My heart sank because I knew I let her down. Then he finished with the most sobering news of all and said, "And your buddy Scott, who was laying down when we left...he broke his neck."

I did not know that, and those last words added tremendous weight to my already guilty heart. After the chat, he was nice enough to drop me off at home because there was no one who could pick me up, then hung around while I took Mom's scorn. When he left, she and I grabbed a cab to the hospital to see Scott. He was in rough shape, but the nerves weren't severed, and so with time he made a full recovery. We're still friends, and can now make light of the accident that should have killed us.

Growing up is hard, no question about that, and with so many close calls born of foolishness, it amazes me that any of us make it through adolescence. But with a bit of luck, a few breaks and hopefully some good mentorship along the way, we do. As I look back on my life, I see that it has been a series of one stumbling block, just like the accident, after another, followed by guidance from someone who cares, and the realization that a lesson learned is part of success.

The accident came at a crossroads in my life, and turning myself in to accept responsibility was the best decision I ever made. Had I kept running, eventually I would have been caught, but by then the charges would have been trumped and would have stuck. And a conviction of any kind would have meant no army service, which would mean I never would have been exposed to the people, places and experiences that shaped my being. Doing the right thing, even after doing the wrong one, doesn't exonerate our actions, but it does keep the doors open moving forward.

Jump ahead nearly a quarter century to Old Quebec and this latest pile up. After exiting Taco to see if anyone was hurt, I was pleased to find only pissed off folks yelling at the young driver at fault. *"Been there,"* I thought with a grin, happy that the only thing he'll have to live with is increased insurance premiums.

When I left the city after my bird-watching, I headed to Montreal for a good explore. But while there I couldn't shake the lethargic feeling I had after the close call. I was coughing because of a sore throat now, the truck needed an oil change, and Athena, as I noted in my journal, "was starting to smell like a barn animal." After two and half months of long days on the road, combined with long runs in the cold, and the pursuit of constant challenges around the continent, I realized all signs were pointing to burn out. But I knew I was in the home stretch now, and so I resolved to keep the pace going strong.

Leg 19
In The Beginning There Was Ontario

"The beginning is the most important part of the work."
- Plato

The wind howled fiercely off the water as I coughed, shivered and waited for the Trolley outside of my Super 8 Hotel in busy Chinatown. That tickle in my throat that had started back in Quebec City, now morphed into a cold that was keeping me up coughing at night. Not that I was doing anything proactive about it, in fact I was still wearing the same jean jacket I had used in the desert, so I had no one to blame but myself.

Still I felt frustrated, and as I buttoned it up, I scorned aloud, "Just buy an effin winter coat already, you idiot." At the slip of the tongue I noticed strangers that were sharing the sidewalk now shuffling away, and grinning at the awkwardness I created, I said as if we were friends, "I've just been too busy to shop, you know."

And it was no lie either, over the last three days I had covered a great distance to get here. Starting in Ottawa, the nation's capital, I enjoyed touring the sites, and was particularly impressed by the exhibits at the War Museum, and a bit surprised too. I didn't realize it held sinister assets like Adolf Hitler's Mercedes Benz. That's the car he famously used to ride around in as his propaganda machine spun Aryan invincibility. It was an artifact that I knew helped to project hatred and injustice, so I wanted nothing to do with it, but I did appreciate the fact it held history.

I visited Parliament Hill, the Senate, House of Commons, and the Peace Tower, then went hopping from pub to pub downtown. Before leaving, I hit up Rideau Canal, and finished by paying homage at the Tomb of The Unknown Soldier. A powerfully poignant place that marks symbolically the lives of over a hundred thousand young Canadians who left to fight tyrants, then never came home.

After Ottawa, I pushed west along the Saint Lawrence as I had since New Brunswick, and passed, via an alluring parkway, hamlets like Mallorytown, Rock Port and Ivy Lea. They were eye candy for the soul. The region is known to be picturesque, and is called the "Thousand Islands." Not that any assistance was required, the entire drive offered spectacular views at every opportunity. And as the name suggests, the swath also contains countless islands strewn about in the middle of the Saint Lawrence.

They ranged in size from big shoals, to small outcrops, and jutted dangerously from the wide channel. Some support entire communities with castle like structures isolated by their water moat and the defences of a dense forest, while others sport modest cabins, or just a single scraggly tree that looks to be rooted, yet floating in the middle of nowhere. What struck most fascinating though was the river's clarity, and more than one local I met told me that in some places, because of the recent invasion of zebra mussels that feed on algae, you can easily see on the river bed shipwrecks in a hundred feet of water.

After that lovely drive, and while freezing my tail off outside from being ridiculously underdressed at the bus stop, I reflected on past legs and was glad to have finally reached the Great Lakes. That is the system of freshwater basins near the middle of the continent, that with rolling tides, fierce storms, their own horizons and one fifth of the globe's fresh water, present as vast as inland seas. Hence the source of the howling wind I mentioned earlier.

Comprised of Lakes Superior, Michigan, Huron, Erie and Ontario, they are the epicenter of a colossus economic zone called the "Megalopolis." The Mega transcends international borders and has a population of some sixty five million. Dotted along the shores of these lakes, and often just opposite from one another in a straight line of sight, are worldly cities like Milwaukee, Chicago, Detroit, Cleveland and where I was today: the centre of the Universe, or so residents here would have me believe, AKA chilly Toronto.

Don't let my good ribbing of the city fool you; the truth is that I love this town. It is after all synonymous with the Canadian image abroad, and typically ranks among the highest in the world for every measurable standard. But to me it feels very different from other Canadian burgs I've been to.

Toronto, or just plain old "T.O." is the largest, most ethnically diverse city in the country, and the fourth largest in North America, so it moves to a pace not felt in realms like Alberta or Nova Scotia. The result of that noise is business, finance, and arts dominance, as well citizenry who sport "big city egos." For instance, back home in Hicksville, when someone says with exasperation that they can't find a "good restaurant in town" or something cosmopolitan like that, already we know they are probably from T.O., because refined folks from the centre of the Universe tend to complain about western cuisine.

It wasn't my first time here by any stretch, I had been many times before, but one particular visit left an impression from when I was young. If you'll recall from the first chapter, it was the road trip here with friends Shuby and Murray, way back in the summer of 91' that was the inspiration for the voyage of discovery I was now on. I was only around twelve at the time, but the randomness I felt then is a force I've been chasing since, and for that, I owe them a debt of gratitude.

They spread the travel bug for sure, but even before the trip, Shuby and I had a bond that was unbreakable. When Mom was dying, she promised to carry the torch of guardian for us, and has filled the role well since. But looking back, the words weren't needed, because she's always stepped up without being prompted. In my many times of need, whether it was being kicked out of the house as a teenager, or broke, down and out, it was Shuby who was there to lend a hand. When we were little, she took us trick or treating on Halloween, even though we didn't have costumes, taught that fishing was an art, not a pastime, and hid clandestine lessons that still unravel today.

I remember the year of the first trip, I needed some spending cash, but I was too young to work. And because Mom didn't do allowances very often, Shuby came up with an ingenious plan to raise funds. Then despite working two jobs at the time, took me out to make it materialize. With a cold call system, and working neighborhoods like grids, she would park at the curb of a house while I went up and knocked on the door. When people answered, I would tell them I was too young to work, needed spending money for a trip, and so had conceived the racket of asking for bottles to recycle.

Some folks would laugh me right off their steps, but most appreciated it as a service and would hand over what bottles they had. Some even gave a five or ten dollar bill and said, "Great ambition kid!" The praise was always a boost, but I would direct the credit to Shuby, and without fail by the end of the night, her old Subaru, which she later gave to my Mom after I crashed the Parisienne, was bursting with loot. We did that for months on end and her gas would have cost more than what I earned.

The generous act showed her integral side, and added another lesson to the pot of lessons she brought to bear. Calm and patient with me always, Shuby was the perfect role model during those unstable years, and I'm certain that if there was a dictionary description and photo beside the terms, "no sweat" and "don't sweat the small stuff," likely it would be one of her smiling face. The drive through Ontario was a constant reminder of how blessed I am to have her in my life.

Flooded with those memories, I was eager to hit some of the spots we frequented back in the day. The first was a short trolley ride down Spadina Avenue to get to Charlotte Street, where I disembarked in the Entertainment District. From there I followed Blue Jays Way to the Rogers Centre where Canada's beloved baseball team plays. Circling the SkyDome, I beamed recalling a memory of Shuby bartering with a scalper for Orioles tickets there.

Next I moved east for a short distance and arrived to the base of the Western Hemisphere's tallest freestanding structure, the CN Tower. When staring up at it from the ground, it looks high, but after an ascent of nearly two thousand feet, via a glass elevator, the scale seems much more intimidating than first thought. But the real test came at the top when it was time to step out onto the glass bottom floor, just approaching it sent goose bumps across my skin.

It was the reason I was there, and I didn't think the heights would bother me after the skydive, but I was wrong. I tackled the fear by shuffling one foot forward at a time, while I used my tippy toes as a prodding device to test the strength of the glass. Then when I was confident it wouldn't bust, I sprinted across, perhaps believing that such nimbleness would see me safely to the other side in the event of a collapse. All it did though was turn heads, and to save face I asked a giggling member of the staff to take my photo, then walked out into the middle to fake bravado and conjured the same rationale as I did in Florida before exiting the plane, "Man I hate heights."

While in the area I headed next door to the "Ripley's Aquarium." It was the first one I've ever been to, so I enjoyed the facility, and was especially intrigued by "The Dangerous Lagoon." It's a long glass tunnel that guests pass through as thousands of aquatic animals, including sharks, sawfish and giant turtles, meander overhead.

And lastly, parched from the walkabout, I crossed over to that inviting building that looks like a warehouse with a fleet of vintage glossy green vehicles parked out front called the "Steam Whistle Brewery." I wasn't really a consumer of the brand before, but after learning the story of the small business, I was, and am now. They produce just one beer, but they do it really really well. It's a Premium Pilsner in a green bottle that was voted tops in Toronto's Golden Tap Awards. Plus it helped earn the brewery accolades as best micro operation in town.

The experience I had there was unexpected and second to none. Because not only was the tasting room a conducive environment to tie one on, it had an open expanse of brick walls, wooden accent beams and a 1960's retro feeling, but the staff were amazing. I had come in just for a quick beer, but because I was the only guy there, somehow I talked myself into a tour.

Which turned out to be a highlight of Toronto that offered plenty of pints, then sent me stumbling out the door.

When I left the city the next morning, I headed south along the 403 Highway through Mississauga, then towards the jettison smokestacks of "Steeltown" Hamilton. The industrial skyline there reminded me of Pittsburgh's blue collar backdrop. Getting lost somehow, I passed through Brampton and found myself in the Ontario Tobacco Belt craving a cigarette. Knowing I had to get out of Dodge before I did something I would regret, I headed back north through Hamilton again, then found the exit I missed. Taking that onto the Queen Elizabeth Way Highway, I settled into a scenic drive through Ontario's Wine Country before ending up in Chippawa, where I got a room for the night.

Awaking early the next morning, thanks to my cough, I knew I shouldn't but I ventured out into the cold anyways for a long run. Leaving the hotel, I planned an out and back jaunt along the Niagara Parkway for a total of eight miles to break in a new pair of shoes I bought in Halifax. During, I picked up a couple of blisters early on, but it felt good to grind out miles in an area I had heard so much of.

Out there, I passed the themed twin cities of Niagara Falls, which looked kinda like Vegas on the Canadian side, but was much smaller. The route was decorated by dry stacked stone walls and attractive older homes overlooking the gorge. About halfway back to my room, I stopped to take advantage of the view at a spot that was only feet from Horseshoe Falls.

Awe-inspiring, the combined Niagara Falls has the highest flow rate of any waterfall on Earth, and the roar of its rush deafened my ears. As the river plummets over the edge, and the water crashes against the rocks, condensation is created, and on this cold morning anyways, it floated up and stuck to everything. The trees, painted white by it, were the most spectacular feature of the frame, because the frost covering them appeared to be like crystals.

The effect of that was amplified as the sun began to rise. Rays refracting through, combined with mist, icy patches on the fall's walls, and the emerald water that moved in a continuous cascade, formed an image that looked right out of a picture book. The site rivaled that of any of the landscapes I had seen yet, and taking it all in, I smiled realizing I was running another of the continent's iconic gems.

The beauty was foreboding too, and knowing it was as much a graveyard as it was a spectacle to share, I closed my eyes to pay homage to all those who have perished here. The list of victims is long and includes daredevils with barrels built like the Titanic, but not quite sturdy enough, tightrope walkers who tempted fate too often, people who committed suicide, and perhaps most tragic of all, those who accidentally fell over the rail after reaching too far.

With all that in mind, I focused my gaze on the froth of brine swirling at its base nearly two hundred feet down, and deduced the pool held more substance than it was letting on. Indeed since I arrived, the entirety of the Niagara Escarpment felt like hallowed ground, and as far as I was concerned, it was a National

Treasure. It was after all the stage where most of the engagements of the War of 1812 were fought.

The War of 1812 was a two and a half year conflict fought primarily by Canadian Militiamen, on behalf of the British Crown, and against Americans who attempted to invade. Best of friends now, it's hard to believe there was animosity once, but the reality is that for centuries the two countries competed for space. The battles that raged here cost thousands of lives, but in the end no lands were exchanged and so both sides claimed victory in a war that saw Toronto sacked, Washington burned, and Detroit captured, then returned.

In that sense it almost seems more like a spat between siblings, than a tit for tat of epic proportions. Our pride comes however from the fact that we duked it out with a country that has ten times our population, and dished back everything we took. In Canada anyways, the war is still taught in schools and as such stirs a tremendous amount of emotion. And it should, because let's face it, even with advanced technology and a sophisticated little military, we don't have the numbers to jump in the ring with the world's superpower, that is, as history records, until our backs are against the wall.

Regardless of who won, and what they won, the actions here did force a beginning. In fact it planted seeds that left Canada with a sense of identity. Instead of being America's 51st State today, we are a proud sovereign territory. Where Gettysburg and Bunker Hill belong to the American conscience, the War of 1812 is a big part of the Canadian experience.

The legacy that exists now between the two neighbors is cordial, but competitive on every issue from sports, to business, to attracting immigrants. In most ways the two cultures are similar, intermixed, and even indistinguishable to outsiders. I don't need to look far to realize that; my sister lives in San Francisco, my brother from a different mother, Billy, went to school in Minnesota, and even my own hometown has a hundred thousand Americans working there. Add to the mix the longest undefended border in the world, the same languages spoken, shared pop culture influences, values, and bloodlines too, and it's clear we have a lot in common. And like buddies who gather for Monday Night Football, we're able to laugh at one another, and pull out the bragging rights card once in a while.

It felt good to run on ground where the nation begun. After the run, I had a hot shower to get feeling back in my numb extremities, grabbed a quick bite to eat, a coffee, then drove to the Rainbow Bridge Customs Crossing with the USA. Entering New York, I headed to Buffalo for the day, but like in Ontario, I wasn't planning on staying. I knew I had a big fish to fry somewhere else.

Leg 20
Midwest, It's Time To Man-Up...

"The man on top of the mountain didn't fall there."
- Vince Lombardi

It was four in the morning as I sat on the edge of the bed in my hotel room with wool socks in one hand and a bottle of water in the other. I was trying to hydrate, but more than that, I was attempting to motivate myself as I watched snow pile up outside. It was a tough sell: I was tired from a restless sleep, which only added to the wayward sense of exhaustion I felt. Plus I was nursing flu like symptoms of a chill, headache, and chest congestion, so was weak and still wheezing with that damn cough of mine.

Things looked daunting, and it seemed like at that moment the Universe was either saying, "Get back to bed you idiot...you're in no condition to run a marathon," or, "You need to finish this trip in epic fashion and man-up, my friend." With a sigh, I hated that

I agreed with the latter, then fired back, "Alright...but just so you know, it's gonna hurt."

After coming all this way, I knew quitting wasn't an option anyways and felt fortunate just to have a spot in the race considering my last minute registration. With my mantra of "explore, experience and push beyond" on the line, I pulled the card I always have and said, "Just get er done...it's one run for the rest of your life buddy." Did I mention yet how much I despise that "rest of your life" part sometimes?

It was February 2nd and day ninety of the trip. After a stint in Ohio to watch a Columbus Blue Jackets hockey game, and a couple of days working the streets of "Motown" Detroit City, I was now in Grand Rapids Michigan to run the Groundhog Day Marathon. The bullet of that big fish needed to be fried, and as you know, had been on my bucket list for some time. In fact, since the trip began I had been trying to catch a race like this one.

In Jackson Mississippi for example I missed the "Blues" run by just days. The same was the case in Charleston South Carolina, then to my dismay, all the way up the Eastern Seaboard. I realized early on that the likelihood of crossing paths with an event in a city at the same time I was wandering through was slim to none. But that's largely because I was unwilling to deviate from my plan of roaming like a nomad, until now that is. The trip was concluding, and there was a sense of urgency to complete this challenge before it was too late.

Sick as a dog though, and on Superbowl Sunday to boot, I dangled the carrot of chili, chips and beers afterward to ensure

I'd tackle the fear of failure head on. I had been running daily of course, but twenty six miles in my condition seemed impractical. Driven by sheer will and irrational determination to get that beer, I pulled my socks on, stood up, and headed out the door.

I was one of the first runners there, which meant I found parking easily, but also that I had to sit in Taco for an hour while pissing in a bottle. You're naive if you think living in a truck for three months doesn't lead to such norms. When people finally did begin gathering at the start line, I made my way over to mingle with the competition.

It was a cold day, but windless and perfect for running I thought. Still I couldn't shake the sense that I was out of place. I'm a minimalist when it comes to running, so for me less is more. Wearing the same red hoodie I used to recreate the "Rocky" saga, and worn shorts pulled over torn leggings that I had for years, I tried my best to play along. About half the racers had familiar garb on like me. But the other half appeared futuristically dressed, and those who had special gear on, had oodles of it. As I observed, I laughed wondering if a similar investment might make me feel as fast as they surely must be.

They wore weird compression socks that were pulled up to their knees, then folded back down, and had layers upon layers of the brightest, moisture wicking material on. For accessories there were electronic gadgets galore, camelbacks with hoses to drink cutting edge formula, and waist belts packed with high calorie snacks that bounced, jiggled and crumpled when they walked.

And even though it was greyer than night by daybreak, some had deeply tinted sunglasses on.

"I'm falling out of touch with this culture," I thought as a guy who was a walking brand passed. Trying to be funny, I quipped, "You look FAST, man!" and got a strange look back. It was a joke, but he didn't laugh, and instead he gave me a furled brow that begged the question, *"Do I know you, bro?"*

"Nope, you don't," I motioned back as I waved him off, then chuckled, "tough crowd."

Somewhere in my lucid state that morning, as I tried to prompt my shoes on, I had hoped the path would be clear of snow by now. But from what I could see of it veering into the woods, it wasn't. Instead it greeted with eight inches of fresh snow that had fallen through the night. I could see "Murph" was offering a baptism by fire for my inaugural flight, but I didn't care, brushing it off, I began to warm up with the astronaut looking bunch.

Once the gun sounded, close to five hundred athletes were off and running in a huff as everyone jockeyed for a good spot. The start looked like a gaggle of uncoordinated extremities, and instantly it was clear that it was going to be a struggle to find our footing. It wasn't long before a single file had developed with the most intrepid warriors breaking trail for the rest of us.

The route repeated like Bill Murray did in its namesake Movie "Groundhog Day." It was six, 4.4 mile loops that cut through the area's generally flat wilderness. With racing I'm usually a slow starter because it takes me a few miles to get my blood pumping, so I used the first lap to find my place in the pack, somewhere in the middle, and tried to come up with a system to tackle the slippery divots. The effort was fruitless however, and as I continued, I was forced to confront the idea that the trail wasn't gonna get any easier.

It took focus to move because each heel strike needed to be found first, and so the trick was to stare two feet in front to anticipate the deep ruts. Which helped initially, but lead to snow blindness, and yup, I was now envious of those damn sunglasses. The combined extra effort it took to make progress felt like I was climbing a never ending sand dune, and as the distance began to rack up, my legs felt like jello and became battered from falling. I was five miles in, and had twenty one left.

Nearing the halfway point, my lungs finally stopped searing from whatever infection was locked inside of them, but now my ankles started to show signs of wear. They were the shock absorbers of the whole deal, and so far had done a great job. But because my feet were sliding into one another repeatedly, the lower tibia's, those are the pointy bones on the inside of the ankles, were banging against each other. After thousands of strides subjected to that, they felt bruised, and looking down I noticed blood was seeping through my socks and shoes. Searching for a silver lining, I began to count miles down, rather than up. I still felt as strong as an ox.

Mile sixteen looked deserted as I arrived at a checkpoint, and asked the girl working the tent if I was all alone. She chuckled thinking I was kidding, then saw my ankles and grimaced before saying, "Do you need help?" I told her no, that I was fine, just a tad thirsty, and needed water to keep going before they swelled up beyond repair. She handed me a cup, and before I left she mentioned most of the runners had already left. Apparently with each loop, because of the trying conditions, they were quitting in droves. *"Interesting,"* I thought, to mock the effectiveness of the expensive gear I saw, then stoked my own ego with their loss. I thanked her, then motored away.

Something happened around mile twenty that hit me like a ton of bricks. I felt defeated, deflated and bested, and realized I had hit that proverbial "wall" runners dread. I couldn't seem to find my second wind to continue, and the struggle to grind out those last few miles now seemed impossible. With nothing left in the tank, I felt like quitting.

I've always prided myself on pushing through adversity, but at that point I saw only what was working against me. The chest cold was nagging, I felt crazy because I hadn't seen a racer in hours, and my whole body ached like an overworked radiator. Stopping to refocus, I leaned forward to catch my breath, and while stretching my spasming hamstrings, questioned, "Why am I here again?" I knew I needed a boost and fast, so I searched my consciousness for inspiration. It was there I found what I was looking for. Smiling now, I began to walk, then trot, then like a locomotive, run towards the finish line as I blurted out, "Dig deep, lads…DIG DEEP!"

As was the case in Oregon during my crisis on the side of the road, it was the voice of Sergeant Holohan that spurred me on.

Midwest It's Time To Man-Up

You'll remember he was that maniacal Drill Instructor who was the "real deal" NCO that used to bark sweet nothings like, "Move expeditiously you thick idiots!" He was as hard as they come, but every once in while he said something fatherly as well.

His endurance was legendary, even by infantry standards, and he would take us on these long morning runs through the woods, which were taxing and would leave ribbons of stragglers in his wake. Then when we got back to the Shacks to cool down, he would pop a cigarette in his mouth, light it, and walk around as we did push-ups, pull-ups and sit-ups, to preach. In a Newfinese accent, he would say, "Your cardio is not that bad fellas...I'll make ya better."

The runs were challenging, but the forced rucksack marches he took us on could make a grown man cry. Forced marches are the bread and butter of an infateer. Physically demanding, they are shuffle like runs over long distances in full fighting order. That is to say while carrying rifles, radios and every other piece of kit the army issued you. The activity was a rite of passage really, and I think he enjoyed them, because he would lead us to a range for the day, then back up a slope we affectionately called, "The Widowmakers Hill."

It was an imposing feature in the training area that came with folklore and the rumour that Holohan had killed soldiers there before just by pushing them too hard up the hill. The rumour seemed like fact back then, but obviously was pure speculation. At worse I had seen guys collapse, but nothing beyond that. Still to collapse of exhaustion isn't normal, and it means a person is pushing with everything they got. Dizzy from not getting enough oxygen to burdened limbs, troops would simply lose consciousness, slump forward, then thunder into the ditch, only to be picked up once the shame truck followed up.

During these death marches, Holohan showed no signs of tiring at all, and even flaunted the fact by running up and down the ranks to torment those who did. He would get right into a man's face, call him weak, ugly, and unfit, then always ask questions that could only be answered with, "YES SERGEANT" or "NO SERGEANT." Of course as young punks struggling to breathe, what we were really thinking was, *"Get lost already!"*

One day I had been limping, like I was on the run now, from a stress injury I had. Seeing the weakness, Holohan was like a shark in the water, and I remember thinking as he charged towards me, *"Ohhh, this is lovely."* Usually he started by threatening to kick guys off course if they appeared to be struggling, but for some reason he grabbed the back of my arm, then marched alongside of me.

"Weird," I thought of the uncomfortableness created by his close proximity. After a minute or so, he said, "What's wrong, Lauritsen?"

I answered back with trepidation in my voice, "Ummm, nothing, Sergeant…just tendinitis in my foot, I've had it for weeks."

Nodding that he knew as much, he said something I could never forget, "Well, it's one hill for the rest of your life, so dig deep okay."

Everything he did was aggressive, and as he slapped my shoulder to make sure I heard, I almost lost my balance. I responded, "Yes Sergeant...I will," then watched him jog away. Once to the front of the column, he turned so we could hear him say over the sound of twenty labouring teenagers, "DIG DEEP LADS...DIG DEEP."

There was something in the way he said it that sent shivers up my spine and made me want to finish what I had started so I didn't disappoint him. But it was his use of term "rest of your life" that resonated so much. First off, it was more emblematic, than accurate, because we did that hill every day in some capacity. But secondly, and in a more broad sense, he was right. He taught that when you're finished a hard challenge, you don't smile a decade down the road remembering the pain, you beam at recalling what you gained. I didn't know it then, but those simple words would define my character until this day.

The fond memory killed that last loop, and as I crossed the finish line with hands held high, I was greeted with a medal placed around my neck, a picture, and the fanfare of just one, a dude they call "Marathon Don." Don Kern is an adventurer, author, and Guinness World Record holder in the sport of running. He was also the organizer of this wonderful event, and I thanked him for that.

Later as we watched Super Bowl together over a couple of Budweiser beers, which gave me an idea, we got to talking. I had been joking about my not so speedy time, which was just plus of six hours, when he cut me off and said, "That's pretty good considering the leaders came in at over five hours because the conditions were so bad, and only thirty nine of you guys finished the full marathon."

In disbelief, his words hit home, and staring back towards the game, I thought, *"Wow…I've finally done the bullet."* The race, which was grueling, was by far the toughest challenge of the trip, but worth every bone crunching stride I took, because forever I'll know I ran a full marathon.

* * *

From Grand Rapids, I headed to Kalamazoo, meandered via secondaries, to Elkhart, then somehow found my way onto Indiana's Route 12. Skirting Lake Michigan as I drove towards Illinois, I crossed the Continental Divide and soon noticed the topography was refining. Where in the days and weeks prior, the ground had often been dominated by dense forests, like the ones I saw along the Saint Lawrence. Now the land rolled gently, and in the same southerly direction I was heading. Opening up as it did, it revealed long lines of sight, littered by farming operations of every kind. I could see I had entered America's Heartland.

It was a picturesque drive and quite the segue to the sprawl of the city on the horizon. In the distance I could see her impressive skyline, which ran a close second to New York City. Smiting that the indulges of deep dish pizza, comedy improv and a culture of Sports Dynasties like, "Da" Bears, Bulls, Cubs, and Hawks awaited me, I patted Athena on the head then said, "Finally sweetheart, we're in the Windy City!"

The entire trip I had been looking forward to visiting Chicago, and so once in town, I didn't waste a single minute. Taking Lake Shore Drive past the industrial might of the South Side, I headed right into the core. There, I parked near Millennium Park, a beautiful twenty five acre green space on the waterfront, and began my tourist jaunt.

Quickly the city reminded me of Toronto in that it sprawls off the water, but the architecture here felt much more complex, and the people more blue collar. Walking towards the ice rink on the north side of the park, which was packed by skaters on this lovely afternoon, I paused to stop and stare.

The vantage overlooked the Magnificent Mile, a strip of Art Deco buildings that reminded me of my stroll down Manhattan's Central Park. But this stretch differed in that it was woven by Neoclassical and Modern elements as well. In one gulp, I pivoted in all directions and could appreciate those influences, and indeed the entire breadth of the Historic Michigan Boulevard District."

Of all the burgs I visited, Chi-town seemed to lead the pack in modern works of art. Opposite the stunning superstructures of Michigan Avenue, many of which have stood for a century, was the Jay Pritzker Pavilion. That's the webbed nest that looks like a dome of steel girders, and that is home to the Grant Park Symphony Orchestra.

Next to that stunning venue, and what I was most fascinated by, was Cloud Gate. Like a moth to a flame, I gravitated its way.

Nicknamed the "Bean" because of its curved shape, it is a highly polished mirrored metal piece that blinds in certain light. Impressed by the way it distorts the skyline, I snapped selfies from every angle.

After leaving the park, I made my way through the grid system of city blocks towards the river and braked at the Western Hemispheres tallest building, the Willis Tower. It's a cool looking mammoth that drew me in and out of the cold for an hour. Following the directions of staff, I went up to the Skydeck on the hundred and third floor. There I got a bird's eye view of the city from the glass ledge and could feel the wind batter the walls. I enjoyed the rush of adrenaline felt, but not wanting to kill the Chicago experience with the same activity I did in Toronto, I headed back down stairs to continue.

From Willis, I moved east via Jackson Boulevard and back towards Millennium Park, but on the way latched onto a tour group that was sauntering through the area. Not totally an ethical thing to do, because I didn't pay for the tour, but they were heading in my direction, so I figured I would follow. You could hardly blame me, the man speaking was like a Wikipedia Page on legs, and as he talked, I became mesmerized by what he was saying.

Using the romance of the Prohibition Era, he began to paint a picture of Old Chicago at a time of tremendous growth. Pointing to the sky he said, "This was a city of cranes then, all working to erect paragons like the Drake Hotel, Tribune Tower and Wrigley Building we saw on The Mile." Then comparing yesteryear to the

Global Hub of today, he continued with, "Sure we're known to produce leaders like Barack Obama and Oprah Winfrey now, but back then we attracted a sinister crowd; this was Gangland, USA."

The town at the time lacked law and order, and so the non-serviced slums, combined with the wealthy innards, made for an ideal melting pot of brothels, speakeasies and gambling dens. With that kind of opportunity everywhere, notorious men like Machine Gun Kelly, Pretty Boy Floyd, and public enemy number one, John Dillinger, flocked here for a piece of the pie. But sitting squarely atop the criminal empire, was the King of Cook County himself, my old buddy, Scarface Al Capone.

I must admit, over the course of the trip, I had developed a bit of a crush on the man. It wasn't by design though, and I had never intended on chasing him like I did Billy the Kid. But for some reason, we just kept bumping into each other. Still leeching off the group, but at a distance, I could hear the guide talking about Capone as he pointed across the street.

Straining over heads, I focused on the prominent landmark he was referencing, it was the Federal Building located at 230 South Dearborn. When he had the group's full attention, he continued, "It's not so recognizable now, because it's a different building, but for those of you who have seen old footage, that's where Al Capone was sentenced."

"*Of course it is,*" I thought with a grin as I conjured the moving picture I had seen a thousand times of him walking up the stairs, then added this spot to the list of places we had already met.

As the guide explained, I followed along with the intent of traveling back to October of 1931. In my head I could see the ghosts of street cars buzzing by and the picture of the burgeoning skyline taking shape. There was an energy about this place, and I could apprehend the Electric Eden of a world Capone once owned. Reflecting back to Alcatraz, I now knew why I saw a sense of loss in his eyes when I visualized him gazing at the bright lights of San Francisco…it was because he lost this, a grand prize as equally beautiful as the city on the Bay. I crossed the street for a better look.

It had been overcast all morning, but now the sun above was burning holes in the industrial haze, and although the vape cleared, a mist remained. I inhaled deeply to see if it was the smell of cigar smoke lingering from the crowd that awaited his arrival. Those who were there, were mostly men dressed in ties, flat caps and their Sunday best. But some brought family members hoping to get a glimpse of the man who runs the Chicago Mob. Part predator, part Prince on any given day, his charm drew people to him.

Just then the crowd hushed, and I noticed a dark coloured Cadillac pull in. "Must be Al," I mused with a smile as the back door, kicked unlatched by a winged tip toe, swung open. Even before his driver could come to a full stop, his burly frame was exiting the vehicle and he was instantly engulfed by the snap, crackle and pop of bursting camera bulbs. When the cloud of the barrage cleared, I could see for the first time why they called him "Snorky."

He looked sharp, and was clean cut, which made the scars on his face obvious. And although he was short, about five feet ten, he had the presence of a ten and half foot man. Wearing his trademark fedora, canted slightly off centre and ready for Florida, he sported a long jacket over a tailored suit that cost more than my house did.

Calm as he approached the doors, I could tell he was making light of this court appearance. His demeanor seemed natural for a man who was round, roly poly, and on top of the world, and he acted like everyone's favourite friend, not the monster his foes avoided. Passing children that wanted autographs, he twinkled their way as if to say, "I'll be right back kid," then disappeared into the courthouse under the guise of another storm of flashing bulbs.

When he left us, he was jovial, but when he returned from inside he wasn't, and as he walked back down the steps, the writing was in his body language. Where only a few hours earlier he was a King, now he was a convicted felon sentenced to eleven years in jail, not for murder or racketeering, but for tax evasion. That seemed like a pivotal moment in time, and as I loitered there to picture it, it dawned that I was standing on hallowed ground. This is where the nation's fascination with outlaws and gangsters ended with the take down of the outfit's most charismatic leader.

I turned around to see that my tour group was walking away without me. Taking the snub as a sign I had overstayed my welcome, I went the other way. I knew I had to close the chapter on Capone, so after leaving the core I drove to the South Side, parked outside his home and covertly snapped some photos. Then I went north to Mount Carmel Cemetery to pay my final respects.

While at his grave, I recounted for him all the places we had met. "I saw you on 'The Rock', at Eastern State Penn, and again in Brooklyn at Coney Island where you were slashed." As I spoke, I wondered why he left such an impression, then deduced that it was because his downfall happened before he could fail. All of Al's successes as a gangster were in his formative years, and because he died so young of dementia, we remember him as he was that day outside of court, the Robin Hood of the Underworld.

In the Midwest I pushed my body harder than I had in years, walked in the footsteps of an American icon, and got to explore cities like Detroit and Chicago. I could have stayed longer, but old hand at the road trip game now, I knew it was time to keep moving on.

Leg 21

The Great Plains Are A Sight To See

"We made too many wrong mistakes." - Yogi Berra

It felt strangely symbolic that I was standing on the Banks of the Mississippi in St. Louis admiring a twenty three foot statue of "Lewis and Clark." They of course are the same duo of daring explorers I had visualized braving the rapids of the Colorado River way back in Arizona. The bronze piece called the "Captains Return" was impressive, and placed here to mark the end of their great Western Expedition. Tired from a life on the road myself, the monument felt like it hinted at the wind down of my trip as well.

It was noon, and after two days of storms that had seen me locked in my hotel room, the clouds were clearing outside, and my cold was finally disappearing. Feeling spry I headed out the door to gallivant downtown then conduct some unfinished business. After leaving the shopping district of Laclede's Landing,

I moved along the rows of giant elms in the Jefferson National Expansion Memorial Park, then stopped to look way up at the beautiful Gateway Arch.

The Arch is a six hundred thirty foot steel leviathan that towers impressively above everything else in the city. It's sumptuous because it's symbolic of St. Louis's place in history. This burg, which was once just an outpost on the fringe of America, grew to become the gateway to the Western Frontier.

After a photo sesh from its base, I headed into the Museum of Westward Expansion, then took the tram to the top for a panoramic. The ride up in the tram was quite confined, uncomfortable even, and with it being only tall enough to sit, the several minute spiral commute felt long and claustrophobic. Once to the observation deck however, the view of the city was awe-inspiring and made it worthwhile. I think of all the monuments I visited on this trip, the Arch was the most magnificent.

Before leaving town, I pursued the unfinished business I alluded to and took advantage of another National Treasure by asking myself, what does Bavarian Pilsner, girls in short shorts, Clydesdale Horses, and the Super Bowl all have in common? The answer was obvious because they're all images of Americana, and synonymous with Budweiser Beer. I don't drink much of it myself, but knowing I wouldn't be back to St. Louis anytime soon, I headed to "The Biergarten" at the Anheuser Busch facility for a quick brewery tour.

Typically I gravitate towards smaller local operations that are either micro, or craft based because when I travel, as with when I'm home, I tend to enjoy the unique flavours of a specific region. For example at home we use crisp mountain water, and home grown Alberta wheat to produce our brew, so it has that distinct Wildrose taste. And of all the provinces' splendid beers, my favourite is a quenching wheat ale called "Grasshopper" made by Big Rock Brewery. So when I hit the pub after work craving a cold one, the first thing I ask for is "a pint of 'Hopper."

It's rare, but if the establishment doesn't carry Big Rock on tap in Calgary, I'll cycle through my other go to drinks: "Velvet Fog," "Blonde," "Wraspberry," and the list goes on. Then, and only then, if I can't find something homegrown I'll resort to what I know is everywhere, "Alright...a Bud it is I guess." It's not that I don't like Budweiser, it's just that it's not as satisfying as Grasshopper, and I know that sounds bias, but broadly speaking, I prefer local fare. That said, I have over the years consumed thousands of "Buds," and because it was local, I was stoked to embrace the "King of Beers" for the afternoon.

Opened in 1852, the brewery is huge and takes up one hundred and forty two acres. Complete with smoke stacks, a clock tower, a packaging plant, horse stables and almost two hundred structures, most of them built of antiqued red brick, it was a joy to walk through, and I'm sure one that would give Buckingham Palace a run for their institutional money. The tour was short, about an hour, but was informative, and finished in fine fashion with two free beers in the tasting room. Now satisfied that I could close the Midwest chapter without having a single regret, I jumped in Taco, drove back through the city, gave the Arch a wave, then ventured into the Western Frontier.

* * *

The weather was mild after I left Missouri, and as I travelled the lonely back roads of Kansas, Nebraska and Iowa, I was romanced by the "fields of dreams" I saw in a sphere that can only be described as "pretty" in nature. Camping along the way again after a trail of cheap hotels needed in the cold north east, I traversed immense swaths of Tornado Alley, which was untouched by urban sprawl and seemingly immune to the hand of time. The expanse was serene and interrupted only by clusters of buildings surrounded by rooted wind breaks of caragana and spruce that signaled the hub of a family enterprise.

What I was witnessing was agricultural compounds that ranged in size, but that had similarities too. Each had at least one quaint house wrapped in a veranda, plus a shop or equipment yard with a combine to work the biggest farms I had seen to date.

I observed that in the yards cats and chickens lived side by side, and free willed dogs chased every truck that passed them by. A lesson not lost on me one frosty morning as I attempted to sneak up and capture a piece of Americana in the form of a brood of crows nesting on a roof. The effort of a photo netted me nothing but a one way ticket, via an undignified sprint, back to Taco's cab. Once there, and safely inside, I rolled down the window, caught my breath, laughed, then mocked the little bastards with, 'I've been chased by bigger and badder dogs on this trip!"

The land they guarded was nearly flat here, but changing rapidly, and was covered in crops like soy, wheat and hops, then cordoned by a flawless grid of gravel roads and ditches overflowing with cattails. Lining those roads was three-strand barbed wired fences, tacked to flush posts pounded deep into the earth. But to add a titch of disorder to the unblemished realm was the odd wind toppled building and set of train tracks running at obtuse angles contrary to symmetry. We were now passing through the Breadbasket of the USA, and trust me when I say, these Great Plains were a sight to see.

It was an immensely reflective leg for me, and for the first time in a long time, communities were separated not by minutes, but by hours that felt like days. As I drove ever deeper into the "Bible Belt," I passed time by having conversations in my head, and chuckled aloud recalling that there were stretches on this trip where I couldn't get a thought in. The talks were two-way, and during them I asked things like, "I wonder if I'm on track in life?" to then follow up with, "And is my calling being recognized?" I think we all have such discussions with ourselves sometimes, but it took me years to realize that what I was actually doing, was praying to God.

I'm sure that by now, most readers have deduced somewhere along this journey of mine, that has included chasing ghosts, visiting cemeteries and conjuring scenes of history, that I'm a bit of spiritual guy. Well it's true, I am, and if you're in my small circle you'll know I don't hide the fact. Like many, I don't preach rhetoric, but that doesn't mean I'm not driven by the virtues of faith in every decision I make. Faith has been a powerful crutch that has seen me through the many highs and lows life tends to throw our way. Throughout them all, I sensed the strength of God was there to pull me through and nurture my soul. That steadfast belief is the reason why I feel like I've never had a hard day and wonder if I ever will.

We all have some higher force we prescribe to, and whether it's a divine one, counsel from a good friend, or a sign from the Universe in general, belief is a powerful engine, and not one to be tinkered with. Respecting that is the reason I don't push mine on anyone else, knowing also that integrity is something we show, not proclaim. I'm happiest when I am helping others, so I believe my calling in life is to humbly serve those in need. To ensure I stay on track and focused on that, I pray, and I do it several times a day.

When I do it's mostly to give thanks, ask for clarity, and wellbeing for family and friends, and it's a practice I've done ever since I was young. But because we didn't go to church, I don't know why or how it came about, just that I've always felt calmest when I speak to God. That prayer doesn't have to be formal either; it could be as simple as asking a question in your head, or giving a kind thought to someone in need. Regardless of the form it takes, each time I ask a question, I get an answer back, and although the answers aren't always what I want to hear, they are guidance nonetheless.

Remember that I'm a list person, and so from those conversations, there were lessons learned and at some point, long ago, I began to record the values I choose to embrace. Because I read them daily, and referenced them often on the trip, I figured I'd have to at least put a summary of them in my book. Here's the Coles notes version of what I awoke to everyday:

Cherish your experiences, both good and bad by seeking out challenges and opportunities instead of passing the buck. Whatever you do in life, whether you're a ditch digger or a CEO, hold your head up high and do it well. Give credit where credit is due, and set your goals high, but when you do accomplish them, don't brag, because accomplishments are like innocence, they don't need to be proved to anyone.

When pushing ahead, reflect, but don't look back because not one of us is a victim on pages that haven't been written yet. Don't be spiteful, the last word isn't vindication, and choose the high road of not hating, even if you're hated. When you see a person in need, be that a homeless person or a colleague, be empathetic, give the benefit of the doubt, then give if you can. And try not to judge others, especially when they ask for help, because it's not easy to ask for anything in life.

Strive to champion tolerance, equality and freedom while sticking to principles of right and wrong, and yes I know, they skirt a fine line sometimes. Be loyal to family, friends and the organizations you serve, because they are responsible for who you are. And most importantly choose to love, laugh, dream, share, explore, experience and forgive those who trespass, that last one is what I struggle with most, but I'm working on it.

Continuing my westerly junket, I reflected on those virtues for days on end while being mesmerized by the constantly changing landscape of the surroundings. From Omaha I went to Sioux Falls and noticed that the cereal crops that had been dominating were transitioning into short grasses as the eeriness of the Badlands began to appear. After passing through a number of tiny towns with names like Scenic, South Dakota, I visited Mount Rushmore and Devil's Tower, then cut back through the Black Hills into Deadwood for a day and a couple of pints of "Pile-O-Dirt" beer at Saloon #10 on Main Street.

There I tilted them back with "Wild Bill Hickok" and joked that I had met his outlaw brothers down in the desert. Well to be fair, it was the chair he was playing cards in when he was shot and killed that I was giving cheers to, not actually Wild Bill. When he fell from that chair he revealed his famed dead man's hand, which foreshadowed his own demise. And take it from locals, if you find yourself holding double aces and double eights, and you're in the company of gunslingers, perhaps start looking for a way to cash out.

After Deadwood I headed south, then spent the better part of the week just meandering infinite leagues from the mile high city of Denver, where I ran the Ralston Creek Half Marathon, all the way north again and into the Montana Territory where I stood on the windswept prairie now. Since leaving St. Louis we had gained

The Great Plains Are A Sight To See

thousands of feet of elevation, and so the thin air here, combined with high altitude dryness, was making my nostrils sting for the first time in months.

I had an enticing view, but I couldn't help but sense that there was loss locked just below the surface. The prairie is static now, but I knew it had an electric side, and visualizing what that would have looked like before the wagons, trains and homesteaders came, the scenery burst to life. Not so long ago this ecosystem would have supported abundance on a scale that rivaled the migrations seen on the Serengeti.

To my front and all around me, the sunbaked grassland would have shimmered with shadows as flocks of birds, the size of thunderheads, passed overhead, and fifty million buffalo sauntered by. Their presence not only tilled the rich black loam, which kept the prairie young, but it also sustained complex Native cultures for several millennia. It seemed like a perfectly balanced relationship between animal, man, and the land that was meant to last until the end of time. But nothing escapes progress, competition, or the profit margin when gold is on the line.

So far on the trip we had delved into the continent's defining conflicts thoroughly. In Quebec we saw that the English ousted the French on the Plains of Abraham. In the former Colonies, we watched upheaval as American Minutemen stunned the world by standing toe to toe with British Regulars in the War of Independence. In Ontario we saw best of friends today blunt one another's ambitions in the War of 1812. And throughout the South we saw the Yankees and Confederates duke it out, at great cost, during the Civil War. Now we were visiting our final battlefield of the trip, and though it may have been smaller than the other ones, it was just as significant. Today we were exploring the Indian Wars, Custer's Last Stand and the Battle of the Little Bighorn.

The Battle, which was just one of many fought during that period, was the culmination of mounting tensions caused by the intrusion of white settlers. When gold was discovered in the Black Hills of South Dakota, prospectors flocked there by the thousands to squat and claim a piece of the pie for themselves. The only problem is that the land was already owned by the Lakota and Cheyenne Plains Indians.

Natives that were already forced onto reservations against their will, felt alienated again, and understandably rebelled to take back what was promised to be theirs. Small skirmishes erupted, which lead to larger ones, and in response the government redrew the boundary lines of the reservation to not include the gold producing regions. Essentially these nomadic hunter gatherers lost their livelihood.

Those who rebelled, left the reservation, or refused to conform to the rules were called renegades and were pursued with the full might of the US Army. With their backs soon against the wall, they rallied behind a great Medicine Man named "Sitting Bull." To unite them he described a vision he had of an impending battle with the Blue Jackets, and guaranteed his followers they would be victorious. After generations of degradation, and even starvation on the reservations, the picture

The Great Plains Are A Sight To See

he painted offered hope, and the warriors of the Cheyenne and Lakota Nations resolved to rise again. The stage was now set for the Battle of The Little Bighorn.

One of the Units dispatched by the government to round up these "free Indians" living outside of the "Rez" boundaries was the 7th Cavalry out of Fort Lincoln commanded by General George Armstrong Custer. Custer was a decorated hero of the Civil War, but an odd icon of American history. He finished last in his class at West Point, had the worst conduct record of any cadet at the Academy, and with scraggly hair, a scrawny build and a moustache to kill, he didn't look much like an officer. Instead he radiated as an eccentric type with a touch of vanity on the side.

He wore a self-styled uniform of buckskins mixed with army issue linen to project the appearance of a frontiersman, and paused for photos everywhere he went. Some praised him as a cunning military tactician, others as a charlatan with too much ambition, but all scholars agree that his risk taking personality was what made him the ideal candidate to quell dissension. It was that reckless side however that would also lead to his downfall.

When reports came in that Indian ponies were seen grazing on the horizon, Custer personally scouted them out, then put together a plan of attack. In his haste to do so, he grossly underestimated the strength of his enemy. I think the Natives probably expected such impulses from him, after all they were wise to the way he used surprise to pillage other villages on the Plains. It might be bold to say, but I'm thinking they weren't so much there to pick a fight, as they were to seek vengeance. The warriors set a trap, and the "Boy General" walked right into it.

What happened next is well documented so there's no need to elaborate with a cliffhanger. Custer, his brothers, and every last one of his men were butchered in the ambush that followed. From the high feature and looking down at the markers placed where members of the 7th fell, it was evident that desperation was the last emotion they felt. The markers grew more frequent with every yard the hill ascended, until at top where a cluster of them confirms the location of his "Last Stand." As I took a minute to contemplate the carnage, I read aloud the inscription on headstones. "U.S. SOLDIER, 7th CAVALRY, FELL HERE, JUNE 25 1876."

It was clearly a decisive victory for the Lakota and Cheyenne just as Sitting Bull had prophesied, but it came at a cost that sealed their fate as well. The US Government responded heavy handed after the massacre, and by autumn most of the Natives involved were forced back onto the reserves, or were fugitives living in Canada. The Indian Wars is a sad chapter of North American history. But the Natives lost more than land in the Indian Wars, they lost most of their spoken languages, their rituals, their independence, and for generations, their proud identities.

The system of reservations that prevails today, in my mind, is segregation. And it is encouraged by the same proxy treaties that created the legacy of racial inequality in the first place. Some circles would even call it "apartheid" or at the very least, cultural genocide. The last battlefield visit reinforced the continent's turbulent past, and that just as the eastern side struggles with the injustices of slavery, we in the west must come to terms with the wrongs perpetrated against Natives.

Leg 22
Coming Full Circle In The Rangeland

"Finding yourself again is as easy as a serene drive."

- A.L.

Continuing west through the Cypress Hills after a brief stop in Regina to tour the Depot of the Royal Canadian Mounted Police, I reveled that I now entered a province in a class of its own. Filled with inviting towns that carry names like Legend, Black Diamond, Picture Butte and Enchant, the countryside was once again transforming before my very eyes; but it would be the last time on this great road trip of mine.

What only a week earlier was the well-defined, nearly perfect, flat prairie pantry of wheat, barley, mustard and hops, now gave way to unruly ranges, ranches, rodeos and evidence of the Marlboro Men who work them best. Arid with olive drab tones, but contoured by water lines and frozen basins of aspen, wild rose, thistle, juniper and native fescue, the land was no longer divvied by barbed wire fences and endless gravel grids, but instead rolled free in every direction with Texas gates, signs that warned of rattlesnakes, and rivers with sandstone coulee banks.

I had been driving the dirt tracks of Colorado, Wyoming and Montana as you know, and had enjoyed filling my photo bank of the vast blue skied expanse the whole time. But all roads seemed to lead inherently here, and to the Crown of the Continent seen jutting off in the distance where refuges like the "Dead Man's Flats" and the "Crowsnest Pass" lie hidden away. If you recall, those are the very same gems the cop who pulled me over in Houston had showed his affection of.

I pulled over to let Athena out and we both stretched our legs with a short walk that was a mile out, then a mile back along a branching dirt path. Once to Taco, I opened the door to let her up, then ventured solo some distance into a field of pronghorn mixed with cattle livestock and sat on a large white erratic rock placed there by a glacier some twelve thousand years ago, then just listened to the Chinook winds blow.

The gusts howled as loudly as they always have, and actually to the point of deafening resolve as if my eardrums were jugs and the wind was the breath of a giant child trying to whisper sweet nothings. With each waft a wave of energy rippled across the ocean of foxtail grass, and in turn the half broken stalks, as rigid as matchsticks, clacked a trillion claps to convince me that I was being applauded, and perhaps even welcomed back.

It was the eve of March now and so the land trickled with meltwater and smelled of half thawed pre spring bliss. It's enchanting intoxication the rival of every kingdom I had visited yet, from the West Coast Rainforests to the Appalachian Frontier, and every nook and cranny from here to there. But not one seemed grander than the other, until just now.

I laughed knowing it was a bias assertion, but couldn't help it and questioned aloud for the wilds to judge themselves. "Is God's Land not the most serene place on Earth?" To me it was, and the way its badlands, peaks, plains and crystal clear waters have always seduced my thoughts with flashes of imagery is proof of that, and gives credence to what a great intrepid explorer once proclaimed right on this very spot, "It doesn't matter how far I stray, I always find my way back to the Alberta I love."

Ok, so notice I didn't say "famous" explorer but rather used the term "great" as a pseudonym for my own expertise as a novice on the subject. As you might have guessed already, the quote is actually just an original "Aaron" but one I use often, and a mantra I live by. In other words, home is where the heart is, and after three months plus on the road, it felt good to be home.

Now I say "novice" in the most benevolent way because I'm not going to pretend for a moment that my footprint will ever compare with the exploits of Captain Cook, Lewis and Clark, or any of the other daring crackerjacks of the New World. In fact, unlike their voyages, mine was no work at all, it was simply a gift and I'm aware the undertaking was epic for just one; me, myself and I. My only real challenge was to accommodate my dog

Athena-bear, which I think you'll agree, was a small price to pay for the comfort of her companionship along the way.

For one hundred glorious days I embraced a spontaneous carefree existence that led to a trip of serendipity. In that time I covered just shy of 30,000 miles via mainly secondary stretches that acted like veins to my addiction of what lie around each bend. I transcended borders, passed through seven of ten Canadian Provinces, thirty eight of the lower forty eight American States, and a countless number of towns, villages, and cities that had more good people than this snapshot could ever convey.

I watched landscapes transition from one extreme to the next, endured tropical heat and sub-arctic blasts, and visited some of the most iconic man-made and natural wonders of the world, then to own them, I ran them all. I confronted historical events, encountered battlefields that forged maps with bayonets, not pens, and spent time chasing the ghosts of outlaws, gangsters, presidents and all manner of influential men, all of whom deserve credit for shaping this unmatched hinterland of ours.

In the different regions I went past comfort by challenging myself to tackle fears while exploring new cultures and experiencing their norms, hence the genesis of my plan from the start which was to "explore, experience, then push beyond." When I was tired I pulled over in rest areas, or slept on beaches. When I wanted a shower, it was cold, or if I needed to do laundry, I rented a room or found a truck stop somewhere. And when I was hungry I hunted whatever it was locals called fine fare. On more than one occasion I paid the piper for that, but usually in the form of an unforgiving hangover.

Now after so much euphoric discovery, the expedition had come full circle and was nearly ending back to where I started from. When I left home the sun was setting, and thinking it would be fitting to conclude by catching the last one of the trip on return, I careened down the Cowboy Trail towards Calgary as briskly as the speed limit would allow me. Once in the city, I weaved through rush hour traffic to get to Memorial Drive, then took Edmonton Trail North to the top of Crescent Heights, where just by chance my favourite bench was abandoned and in wait.

Sitting there I celebrated with a deep breath of fresh air and laughed that I had made it just under the wire. It was chilly, but easy to relax, and in no time I was admiring the city nestled in the Bow River Valley as the windows of her many skyscraper's began to glow warm with a spectrum of orange and purple that proves the sun's duress is always brightest just before dusk. With the familiar mountain backdrop to spur my thoughts and thanks, I reflected on how many places around the continent I had taken time to sit on my tailgate, pray, crack a beer, and watch this same thing.

Too many to count I rationed with a grin, but focused on a few that can't be undone. There were the couple of evenings I sat perched on the seaside cliffs of California teetering on the edge of nothingness. Then there was the intense fuchsia that livened the desolate desert of the Southwest, then segued into a black starry night. And of course I would be remiss if I didn't mention the silhouette of Manhattan's jettison skyline plastered against a red abyss.

Then reflecting further I laughed with revere at how often I got lost but didn't care, snuck Athena into somewhere, or locked my keys in Taco then cursed to the ire of a spectator, "ENOUGH ALREADY, MURPH!" The notions continued to pour and I wondered how many nights I laid awake to listen to coyotes sing, or drifted to the patter of a rainstorm, then awoke startled remembering my shoes were out there.

Not everything was golden though; there were occasional frustrations and disappointments too, especially if I missed a goal after a long haul. And I know I should have mentioned some hardships throughout, but complaining or dwelling on adversity isn't something I like to do, and so they're words I choose to rarely use. Sour days don't get a say in my chronicles of the road.

This book is my first attempt at intellect, and really the first of anything I've written since I grappled with grade eleven English. So I know it won't appeal to all because it doesn't follow the rules of literary law, but it is told the only way I know how, just as I think it in my head.

I never even considered the written expression as a course, until the beautiful woman in my life encouraged me that this was a story that had to be told. She planted the seeds, and when I mocked my own ability with, "A book? What book, I don't even read! Where would I start?" She fired back, "Just write it, love." And so I did, I pushed aside time reserved for friends, family and even personal ambitions to immerse myself in my mind and just write this thing.

And I am grateful for the push and her show of confidence, because like the journey itself, the book was therapeutic with highs and lows, and although the open road closed a lot of chapters for me, the book conjured just as many that I wish I could forget. But such is life, and I've been blessed by every of those building blocks, even if they linger in the form of bad memories from my past.

The brilliance of the trek was unexpected but blatantly clear. The further I delved into the continent's boundless spaces, the more parallels I drew to my own places and the more reflective I grew with reconciliation. On the road, as in life, I've always believed that when you don't know where you're going, any road will take you there, so just because I didn't follow the crowd doesn't mean I wasn't blessed at every wrong turn.

Nearing the end of the trip, and aided by the vastness of the west, I realized that through my life I had racked up a diverse set of experiences, and deduced also that somewhere along the line I had entered the sovereign of worldly man. I just didn't know how, where or when. Smiling at the concession, I know the young Aaron from twenty years ago would be envious because it was a promise he aspired to fulfil. Though I'll remind him, my inner self that is, that we still have a long way to go.

I had set out to find myself again, but I ended up with a lot more than I bargained for, I found my second awakening. So this isn't a typical travel guide in that I don't detail the location of good parking, slummy hotels or tell you that a meal was a tad cold, because that stuff doesn't matter to me. My only hope is that some will embrace the writing style, and those who do will curl up with a glass of red wine or a bottle of beer somewhere to enjoy the ride as much as I did, or perhaps even take the plunge themselves.

But to satisfy the travel gurus who do appreciate a critical review, I'll offer some insight on a few things that caught my eye. But beware, the nonspecific categories from my own junket might be a bit broad to plan anything other than a serendipitous trip.

With that in mind, here goes;

    1) Best Brewery - Steam Whistle in Toronto, Anheuser Busch in St. Louis and Keith's in Halifax.

    2) Best Pub Experience - The Lower Deck in Halifax, The London Grill in Philly and McSorley's in New York City. All of which are on the East Coast, no surprise there.

    3) Top 5 Skylines - New York City, Chicago, Vancouver, Toronto and San Francisco.

    4) Best Beaches - Cannon Beach in Oregon, Long Beach in BC and Far Beach in Florida.

    5) Nicest People - ???, Even if I took out friends and family there would be too many to count, so I'll go with random strangers, Michael and the Birdman in NYC.

6) Best Runs - Wildcat Trail in Arizona, Rainforest Trail in BC, Central Park in NYC and of course the snow laden marathon loops of Grand Rapids Michigan.

7) Best Adventure - SkyDive in Florida, Surfing in British Columbia and Saturday Night Live in New York City.

8) Best Drives - Tough one but, 101 Oregon Coast, Pacific Rim on Vancouver Island, Route 66 leg in New Mexico, and the Bow Valley Trail in Alberta.

9) Best Landscapes to Watch Sunsets - Alberta, Arizona, New Mexico, California, and Kansas.

10) Most Compelling Cemeteries - Arlington in DC, Saint Louis No.1 in New Orleans and Granary Burying Ground in Boston.

11) Most Historical Communities to Walk - Old Quebec City, New Orleans, Boston and Philly.

12) Funnest Ghosts to Chase - President Kennedy, Martin Luther King, Billy the Kid and Al Capone.

13) Most Haunted Places - Eastern State Penitentiary in Philadelphia, Alcatraz in San Francisco and Salem Commons in Massachusetts.

14) Most Impressive Monuments - Washington DC in general, Mount Rushmore in South Dakota, and the Gateway Arch in St. Louis.

15) Most Walkable Battlefield - Chattanooga on the border of Tennessee and Georgia, Little Bighorn in Montana and the Plains of Abraham in Quebec City.

I could spin a thousand categories and they still would barely scratch the surface of the continent, so the above might not help much, but if I tried to deviate too much from the way I wrote in order to be a tour guide, the project would have felt like work. So I'm glad I didn't because to get it finished was already a haul that took longer than I expected. Hats off to professional writers for their resiliency, but the experience was a joy and a privilege and now one I know takes a village.

Imagine what resources it consumes for a grown man like me to take a two year hiatus from work and the real world to write a book; take my word that it's a lot and that I couldn't have done any of it without the support from family and friends. This book is as much yours as it is mine, I want you all to know that. And sorry for the reclusiveness these last twenty two months, I learned early on that to be able to reach my full potential, I needed to hide my phone away. God knows that for every one page published, I had to write ten then cut the fat. Which took time, energy, and a whole lot of fourteen hour days.

People say I'm lucky to have gone in the first place, and feeling fortunate but never once privy to a silver spoon, I agree, but remind that "auspicious" is a more appropriate word. We make our own luck in life, and the truth is that I willed the trip while at a crossroads that included a divorce and the sale of my business. It wasn't cheap though, and at a cost of a little over a dollar a mile in food, gas, accommodation and not one adventure spared, the price did add up fast.

To pay for it, I used savings from an adulthood of laying bricks as a landscaper and from soldiering in the far corners of world. So every penny was hard earned through blood, sweat and tears and I wouldn't have had it any other way. I know that if the money wasn't back breaking to earn, the experience wouldn't have the same value, nor would it have capped years of struggle, so I make no apologies for being "lucky" in any way.

Shuby said it best before I left with, "The experience of travel pays more in life dividends than any bank could pay in interest." It was sound advice from a guardian I trust, just add it to that overflowing bucket of wisdom she's given me over the years. After a decade plus of talking about putting my worries behind me in favour of something freeing like a road trip to hit the reset button, I finally manned up by tying loose ends, putting a tenant in my house, and just doing it.

Since I got back a lot has changed. My good friend, and Shuby's soulmate, Murray, has passed away. God bless him, and I know he was reading the book from over my shoulder as I wrote it. So if you didn't like it Murray, save your comments for when I see ya again, which hopefully won't be for about fifty years, no offense buddy. As for me, I'm shacked up with that wonderful woman I courted not long after my return. The very same lady who encouraged me to write this thing and who I proposed to in dedication. If you've read this far you might be curious to know Valley said yes to me and that were over the moon because our first child is on the way! The future looks brighter than bright each day.

People ask, "What travels lie ahead Aaron? Will it be a hundred days in Asia, Europe, Australia, or somewhere we've never heard of?" I respond affably so as not to give too much away, "Time will tell...but something's brewing, and trust that it'll be one hell of a journey!" Until then my friends, dream big, be random, seek chance opportunity, and don't be afraid to put your problems in the rearview mirror like I did to "explore, experience, then push beyond" with your own liberating version of The Great North American Road Trip.

Drive Safe And Have Fun!!!